SURMOUNT STYLE + COPY

突破风格与复制

MID-RISE/MULTI-FLOORS
小高层 / 多层

杨锋 主编

江苏人民出版社

PREFACE

序 言

REVELATIONAL RESIDENTIAL DESIGNS

Demographic patterns and social structures influence today's travel, living and buying experiences. Living life to the maximum means that the dynamic of changes and trends impacting our daily lives shows the world in a state of constant transition.

Society enjoys a seamless blend of both sensory aesthetics and functionality in all residential arenas. Innovative residential design stems from conscious efforts to create a sustainable living environment of unrivalled quality. Designers strive to ensure that delicate balance is fused with current and future needs of the appropriate target audience.

For each individual project, dwp believes that architects should make a thorough assessment of the client needs, the targeted potential buyers and dwellers, along with all relevant environmental, cultural, emotional and spiritual aspirations and practical concerns, to be factored into the equation.

The process of innovative residential is begin from conceptualizes, plans to tailors into residential designs, to best suit the individual requirements and lifestyles of their clients, neither compromising on aesthetics nor functionality, in order to achieve optimum results, stellar designs and residential designs that exceed client expectations.

Using extensive knowledge and industry know-how, designers should have to entirely dedicate to creating iconic residential design solutions from masterplanning, architecture, landscape, through to stunning interiors, with a client-oriented focus, making ideas and dreams into reality.

Reflecting the diverse lifestyles, cultures and cities, dwp has conceived and implemented inspirational, award-winning projects. EverRich II and maysan are two of excellent innovative residential project by dwp, which explains perfectly about dwp's understanding of innovative residence. Maysan Towers' design theme draws influence from the unique natural topography of Al Ain, often referred to as the Garden City of the Gulf, integrating into nature harmoniously. While the materials and finishes in EverRich II are in warm tone, which put more focus into a feeling of home to create a cozy, intimate feeling, with a touch of luxury and elegance.

This is a precious innovation in architecture; it may not be the best but it reflects dwp's creative willingness.

Brenton Mauriello
dwp Group CEO

创新住宅设计的启示之声

人口模式和社会结构影响着人们的旅行、生活和购物体验。生命在最大程度上意味着各种变化和趋势的不定性，影响着我们的日常生活，显示出世界正处于一个不断转变的状态之中。

在所有的住宅领域里，美学和功能的完美融合都深受人们的赞赏。创新的住宅设计源于有意识的努力，这种努力旨在创造一个最高质量的可持续生活环境。设计师们力求在满足目标群体的当前需求及未来需求之间保持微妙的平衡。

在 dwp 的设计理念里，建筑师们理应对每个项目都先做好全面深入的评估，包括客户的需求、潜在的目标买家和住户，然后把这些因素与相关的环境、文化、情感和精神的诉求，以及实际可行性综合起来进行全方位的考虑。

设计创新住宅的过程开始于一系列概念化的过程，也就是为住宅设计量身定做规划图纸的过程，以更好地满足客户的个性化需求和生活方式。当然，为了最终做出超越客户期待的一流住宅设计，达到最佳设计效果，这一过程不会向美学或功能的任何一方妥协。

从总体规划、建筑风格、景观配置到极富魅力的内饰装潢，设计师应该使用广泛的知识和专业技能，全心全意地专注于创造具标志性的住宅设计方案，以客户导向为重点，让概念和梦想成为现实。

我们也设计和实现了许多鼓舞人心的获奖项目，这些项目反映出不同的生活、文化和城市风貌。其中越南胡志明市 EverRich II 公寓和阿联酋阿布扎比市米桑大厦就是我们两个杰出的创新住宅项目，完美诠释了我们对创新住宅的理解。米桑大厦的设计主题源于有"海湾花园城市"之称的阿莱茵市独特的自然地形，与自然和谐地融合在一起。而 EverRich II 公寓的材料和装饰则把更多的重点放到居家的感觉上，采用了暖色调来创造一种舒适、亲密的感觉，并透出淡淡的奢华与优雅。

在建筑设计上，这是难能可贵的创新，它不一定就是最好的，但它反映出我们的创新精神。

布伦顿 • 毛列洛

dwp 集团 CEO

PREFACE
序 言

What's the latest residential design in western developed countries and which direction they are moving forward when many residential designers are busy copying indiscriminately Italian, French, British and American styles?

China is one of the world's largest housing market, however, the majority of residential designs still remain on copying styles, which such indiscriminately pursuit worldwide styles and reproduce mechanically large amount of living spaces. It is a major problem for Chinese contemporary residential design. Indeed, designers have realized that it will never be a scientific development with such so-called popular styles originated from blind pursuit and self-respect of customers. Actually, personal identity and status are not only recognized through the Beijing Tuscany Style. With the growing confidence and more internationalized perspective, residential designers shall break the replication of copying styles and be more rational about "man and nature" so as to be people-oriented and fulfill environmental demands and to ensure that the residences in China will be in a healthier development. Moverover, the growing imperatives for sustainability, shortage of land, ever-changing modern technology and architectural materials enable designers to create a more comfortable living environment needed in the modern society.

To fix the crux of residences in China and provide a new thinking style for designers to break the replication of copying styles, we collect 84 world's typical residential designs such as high-end residences in international metropolitans like New York, London, Paris, Tokyo, etc., along with mid-range, economic residences in developed countries to fully demonstrate the latest trend in the world. These projects are unique architecture with harmony between form and function, not only focusing on daylighting, views, ventilation, public and exterior spaces of residences, but manifesting the modern residential design trend is towards green, ecology, intensivism, science and technology, reflecting diversity, multeity and organism of architectural concept elements. The book summarizes and analyzes each project to bring a fresh thinking about the most advanced residential designs (totally 84) in the world for residential designers lingering in the replication of copying styles, helping them break it out, fully realize people-oriented and harmoniously integrate man into nature.

HongKong Architecture Science Press
Editorial Board

当国内住宅设计师盲目抄袭意、法、英、美式风格时，这些西方发达国家的最新住宅又是什么样的呢？他们的设计在朝哪个方向发展？

中国有世界上最大的住宅市场，但住宅设计多数仍停留在对风格的抄袭上。盲目追求世界风格，机械地大规模复制生活空间，是当下中国住宅发展面临的一个重要问题。设计师们知道风格的盛行很大一部分源自消费者的盲目媚外和自我尊重需求，实际上身份与地位的认同并非一定要用北京的托思卡纳风格来实现。伴随着国人自信的回归和越来越国际化的视野，国内住宅设计建筑师只有跳出风格与复制的怪圈，才能更理性地思考"环境与人"的问题，中国住宅建筑才能更健康地发展。而日益增强的可持续发展要求、居住用地的紧张，还有现代科技和建筑材料的日新月异等因素，使得现在的设计师有能力也有需要营造更舒适的人居环境。

针对中国住宅的这一症结，为了给设计师走出风格与复制的怪圈提供新的思路，本书收集了 84 个世界优秀集合住宅设计典范——既有发达国家的国际大都市如纽约、伦敦、巴黎、东京的高端住宅，也有它们的中档和经济住宅，全面展示了世界住宅的最新动态。它们都是形式与功能和谐统一的个性化建筑，除了住宅建筑需要侧重考虑的日照采光、视线、通风、公共与户外空间等因素外，它们还反映出当今住宅设计的趋势和方向是绿色、生态、集约、科技，突显着建筑理念的多元化、多重性和有机性。本书对每个项目的特点进行了总结归纳和评论分析，为国内徘徊在风格与复制中的建筑师们带来 84 种世界最先进住宅设计的全新思维，帮助设计师突破风格与复制，真正做到"以人为本，天人合一"。

香港建筑科学出版社
编委会

CONTENTS

目 录

CONTENTS

目 录

Parc Marianne Residential Building, Montpellier, France

Location: Montpellier, France
Architect: AS.Architecture-Studio
Gross Floor Area: 3,160 m^2
Number of Floors: 5
Photographer: BENOÎT WEHRLÉ

法国蒙彼利埃市 玛丽亚娜公园住宅楼

地点：法国蒙彼利埃
设计公司：法国 AS 建筑设计工作室
建筑面积：3 160 m^2
层数：5
摄影师：贝努瓦 · 维尔勒

Surmount 突破点

FACADE 立面

SHADE 遮阳

ENERGY SAVING 节能

MATERIAL 材料

AS. Architecture-Studio will design 32 flats shared out into 2 apartment buildings made up of 5 floors each. They will be connected through a common underground car park area.

This project consists of 2 apartment buildings of 16 flats each and a common underground car park area.

The Municipality of Montpellier has decided to balance the land use of the Port Marianne mixed development zone within the town's borders. They also aim at extending the Antigone urbanization towards the seaside.

The building base has a minimum height of 5 m, is made up of raw concrete and will link the 2 buildings together. It will also create some kind of unity inside the area and bear the entrance halls. The setbacks of the first and fifth floors break the façade line; a continuous balcony is thus created at these levels. Architects have also intended to offer wide views onto Parc Marianne through large terraces.

Both buildings must emphasize green areas and adapt to their immediate environment, as they are surrounded by a park and the Lirode gardens. Thus, AS. Architecture-Studio has taken into account the eco-friendly approach when choosing materials. These buildings have natural colour wooden façades which suitably integrate into the mixed development area's style. The setbacks of the first and fifth floors are tin-made. These floors are thus different from the rest of the building and benefit from natural protection against sunrays.

Site Plan 区位图

法国 AS 建筑设计工作室将设计两幢包括 32 个住宅单位的住宅楼，住宅楼各有 5 层高，由一个公共地下停车场连接。

本案为两幢各自包含 16 套住宅的住宅楼和一个公共地下停车场。

蒙彼利埃市政府正在着手进行玛丽亚娜港口综合开发区空间的重新规划，以延长安提戈涅城市的海滨线。

至少 5 m 高的混凝土基座把这两幢住宅楼连接到一起，将整个住宅区的内部连为一体，并容纳了入口大厅。建筑主体立面被首层和第 5 层的回缩所中断，因此从第 2 层到第 4 层形成了连贯的外延式阳台。建筑师通过打造大阳台，让居住者可以欣赏到玛丽亚娜公园区开阔的景观。

位于公园内的这两幢住宅楼在突显它优越的周围环境的同时，更要加强自身对自然环境的保护。所以，法国 AS 建筑设计工作室选择的材料都是生态友好型材料。建筑物的立面都选用了原色的木材，与综合开发区的环境和谐地融合在一起。首层和第 5 层的回缩表面由锌包裹着，不但显示了它们与其他楼层的不同之处，还避免了阳光对建筑的直接照射。

Master Plan 总体规划图

PARK AVENUE

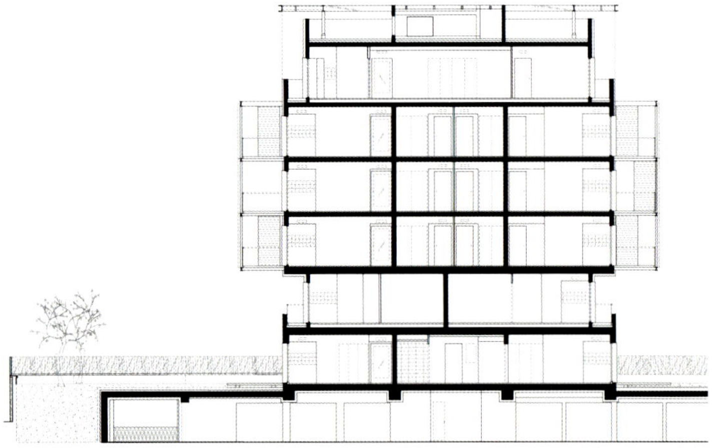

Section Plan 剖面图

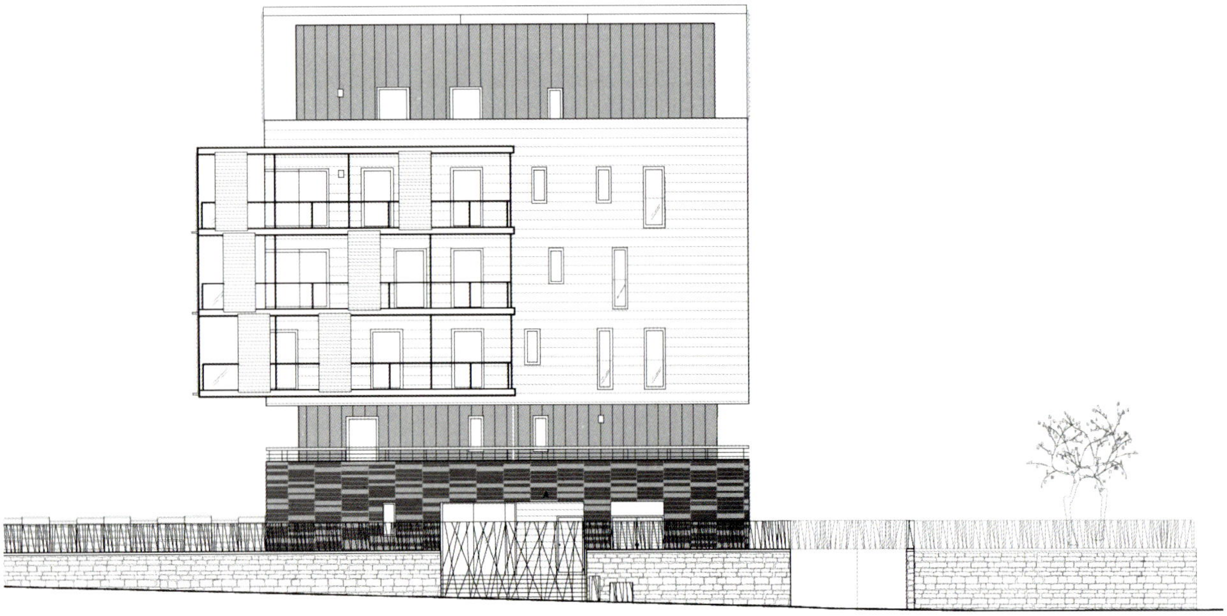

North Elevation Plan 北立面图

FEATURE 特点分析

Shade

A pergola protects the southern façade against sunrays and then bents onto the roofing. Not only does the pergola reduce the solar impact on both buildings, but it also contributes to the creation of a common style for this mixed development area.

遮阳

为了避免南面强烈的阳光照射，从建筑的南立面到楼顶都被一个巨大的凉亭保护着。这个凉亭不仅减少了阳光对建筑的影响，还为这个综合开发区创造了统一的风格。

East Elevation Plan 东立面图

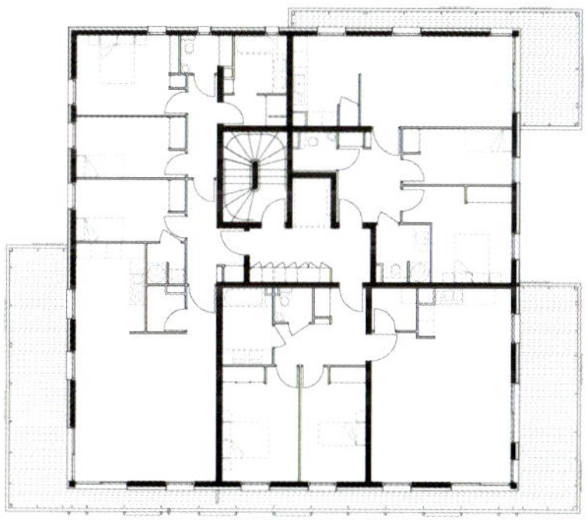

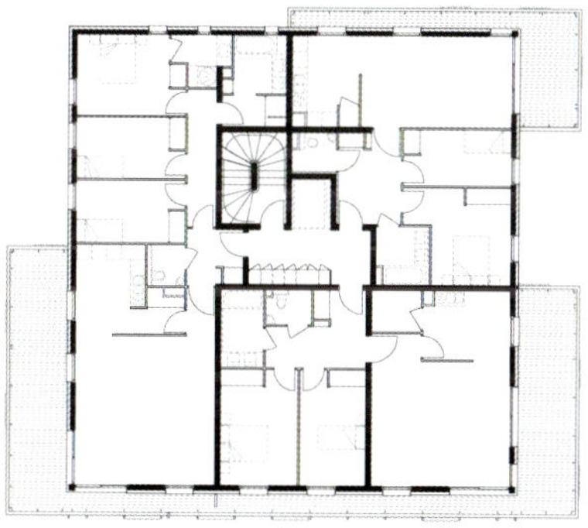

3rd-5th Floor Plan 3~5 层平面图

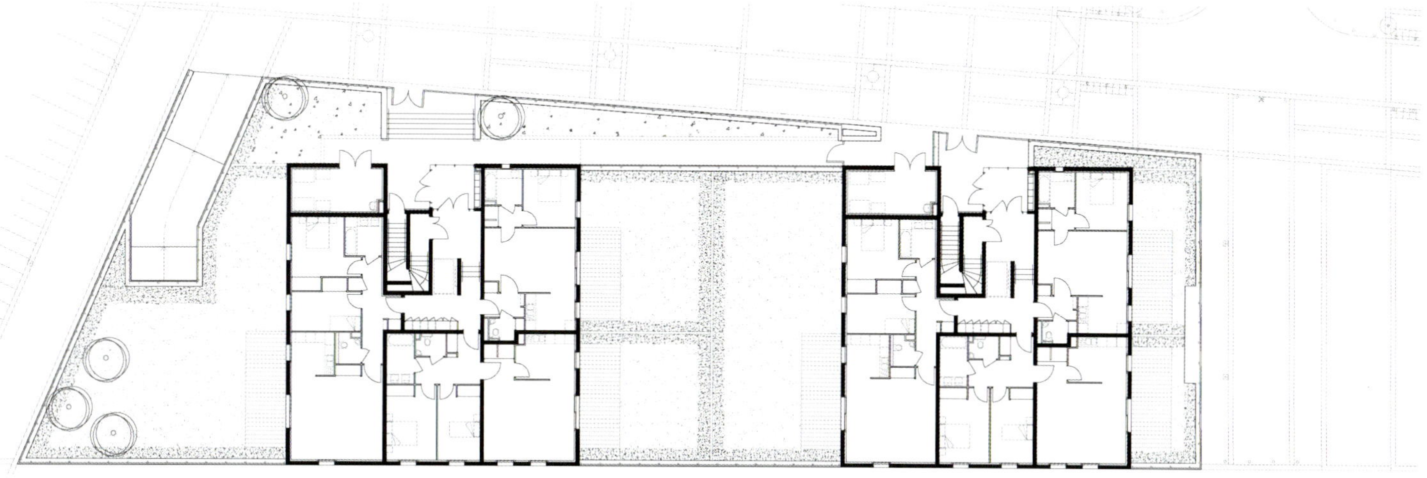

1st Floor Plan 1 层平面图

Lyon Confluence, France

Location: Lyon, France
Architect: Massimiliano and Doriana Fuksas
Gross Floor Area: 28,955 m^2(Residences); 1,807 m^2(Commerce)
Number of Floors: 10
Photographer: Philippe Ruault

法国里昂汇流街区

地点：法国里昂市
设计公司：马希米亚诺与多丽安娜·福克萨斯建筑事务所
建筑面积：28 955 m^2(住宅楼);1 807 m^2(商务楼)
层数：10
摄影师：菲利浦·胡欧特

Surmount 突破点

FACADE 立面

SHAPE 造型

MATERIAL 材料

DAYLIGHTING 采光

The Confluence, well before the great and important planned infrastructures, represents a place of strong character, a beautiful natural setting, with the green hill of Sainte-Foy, the two rivers and the proximity of the historic center of Lyon, full of history.

The project starts with a study around the possibility of building a housing complex around a large void – the park – that opens into another void – the dock. The two voids, the park and the dock, constitute the skeleton of the project. To locate the volumes, priority has been given to the longitudinal direction of the park and the dock. The idea that brought to build another void between the buildings is the heart of the area. The central part was left open to the west side, looking to the hill.

The voids between the buildings are never in line and offer different views towards the park and the docks. The construction of the volumes is entirely based on the interstices where the light of the voids delimits the solids.

The buildings on the dock were associated to the realistic and poetic image of the cranes/containers, as a reminder of a fully operational river port. A piece of city is reflected in the water with all its colors, with all its different materials, creating a multitude of vibrations. The building close to the park and towards the Rue Denuzière opens its volumes like a kind of drawers, opened to the warm sun of the West and toward the green hill of Sainte-Foy. Everything participates to the will of city in movement, always in progress; everything seems like playing with the light that hides, highlights, and modifies the different architectures.

汇流街区曾经是极其重要的规划建设项目，毗邻充满历史气息的里昂市中心，它代表了该地区鲜明的特征，拥有美丽的自然环境，包括葱绿的圣弗伊山和两条相汇的河流。

经过研究，项目计划建造一栋综合住宅楼，周边的空地是一个公园，可以通向另一片空地——码头。这两片空地——公园和码头，组成了该项目的主框架。建筑的定位优先考虑公园和码头的纵向方位，而将建筑群之间留出的一块空地作为地带的中心。中央部分向西侧敞开，以便使人欣赏到山丘。

建筑群之间的空地是非线性的，可以让人从不同角度欣赏到公园和码头。建筑的体量结构完全建立在空地日照间隙所界定的范围之上。

这栋码头上的建筑与起重机和集装箱组成的现实而诗意的画面联系起来，提醒着人们这是一个全面运作的港口。城市的一角倒映在水中，其斑斓的色彩、不同的材料，在微波荡漾中呈现出多样的形态。接近公园朝向德鲁热大街的一个建筑体量呈敞开的姿态，像某种抽屉的形状，沐浴在西部温暖的阳光下，面向葱郁的圣弗伊山。一切都在运动中参与到城市的发展愿景中，不断地进步；一切似乎都在和或明或暗的光线嬉戏，光线突出了并修饰着不同的建筑。

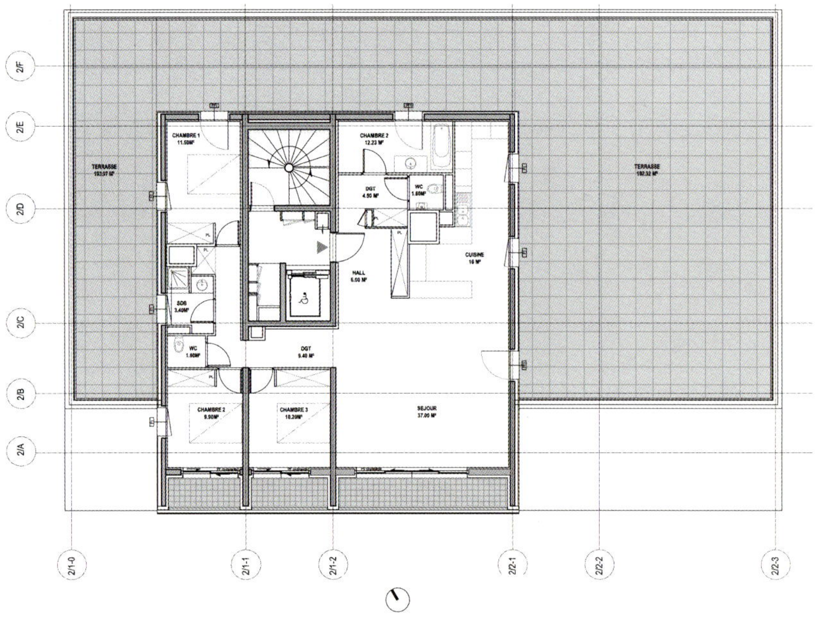

Floor Plan　楼层平面图

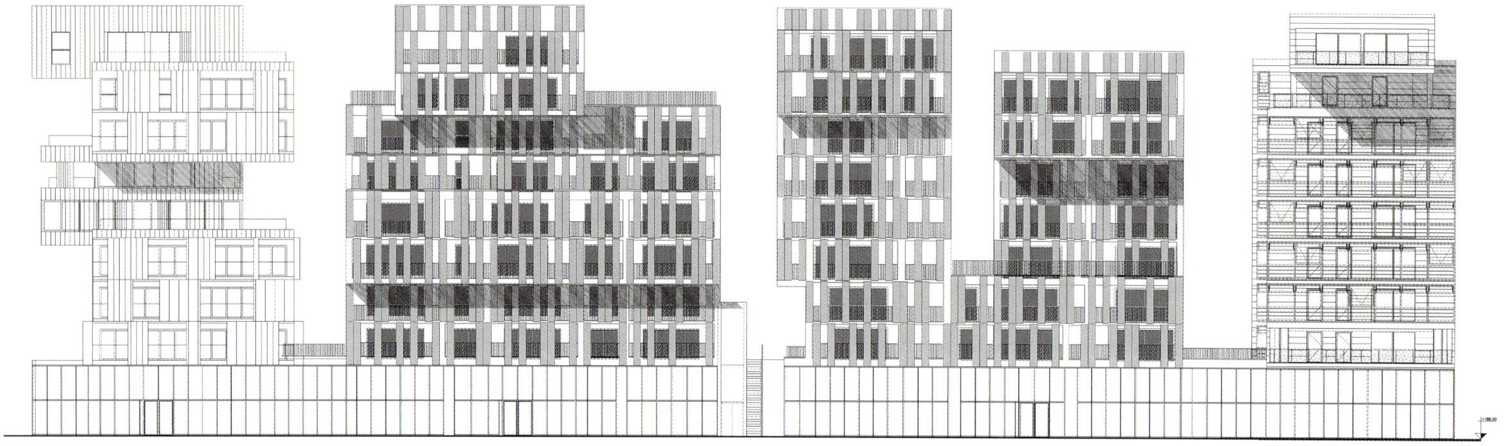

Elevation Plan 立面图

FEATURE 特点分析

Shape

On the volumes of the buildings, in a continuous play of light, a series of rotations and distortions, give life to blocks of matter that become habitable sculptures.

造型

建筑体量在光影的连续不断的变化中呈现一系列的旋转和扭曲，赋予了建筑生命，使其成为一栋雕塑般的住宅楼。

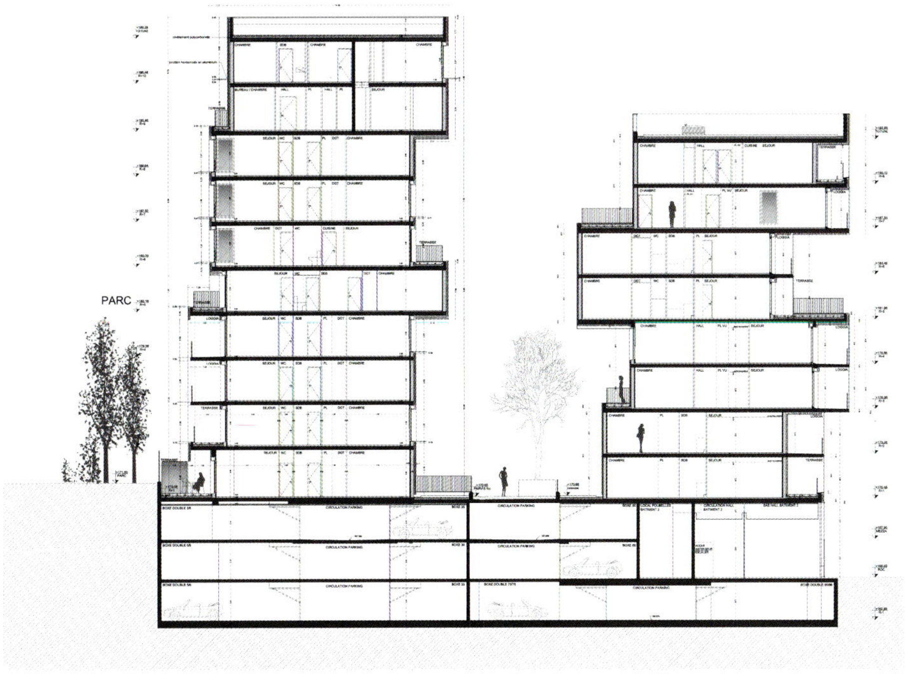

Transversal Section 1　横向剖面图 1

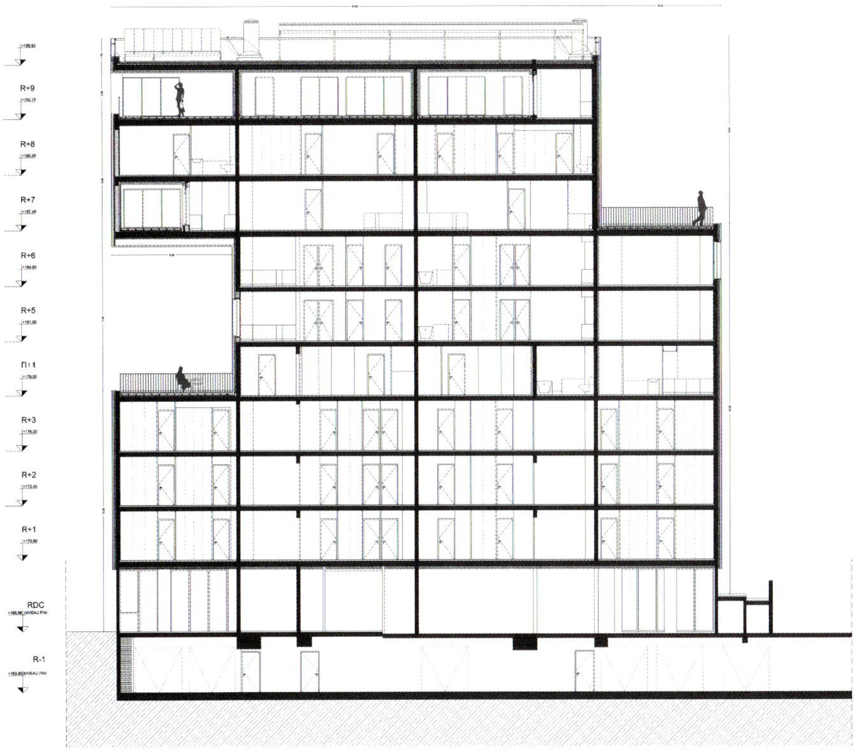

Longitudinal Section　纵向剖面图

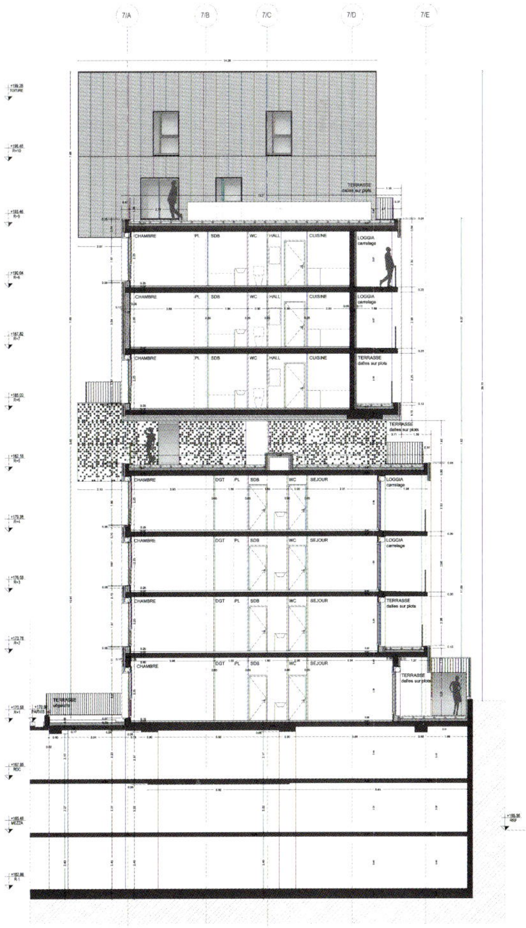

Transversal Section 2　横向剖面图 2

Metropolis Apartments Copenhagen, Denmark

Location: Copenhagen,Denmark
Architect: Danielsen Arkitekter A/S & Future Systems
Site Area: 1,020 m^2
Gross Floor Area: 13,800 m^2
Height: 42.2 m
Number of Floors: 11

哥本哈根城市公寓楼

地点：丹麦哥本哈根
设计公司：A/S 丹尼尔森建筑事务所、未来系统事务所
占地面积：1 020 m^2
建筑面积：13 800 m^2
高度：42.2 m
层数：11

Surmount 突破点

SHAPE 造型

FACADE 立面

BALCONY 阳台

MATERIAL 材料

The residential tower "Metropolis" is a new spectacular residential development. It is located on a small peninsula called Sluseholmen in the Copenhagen Harbour. It is possible to sail right to the front door of "Metropolis". The area has historically been one of primarily industry and factories, but today the area is one of the most progressive residential developments in central Copenhagen. It is a beautiful area with a pleasant atmosphere inspired by residential neighbourhoods along the canals in Amsterdam. The gentrification of the area has meant that a wide array of new shops and cafes as well as many multinational companies has established themselves in the area.

The Danish Architectural firm Kasper Danielsen Arkitekter A/S in collaboration with the internationally renowned firm Future Systems, is behind the design. "Metropolis" is a 11 storey all residential development. The 81 condominiums vary between 90 m^2 and 210 m^2 and all units feature minimum 2 balconies. Traditionally the bridges in the harbour have featured green cobber clad towers. Together with the industrial silos these towers has served as an inspirational motive for the design of "Metropolis". The projects curved forms is clad in fiber-concrete shells in a blueish-green colour, all

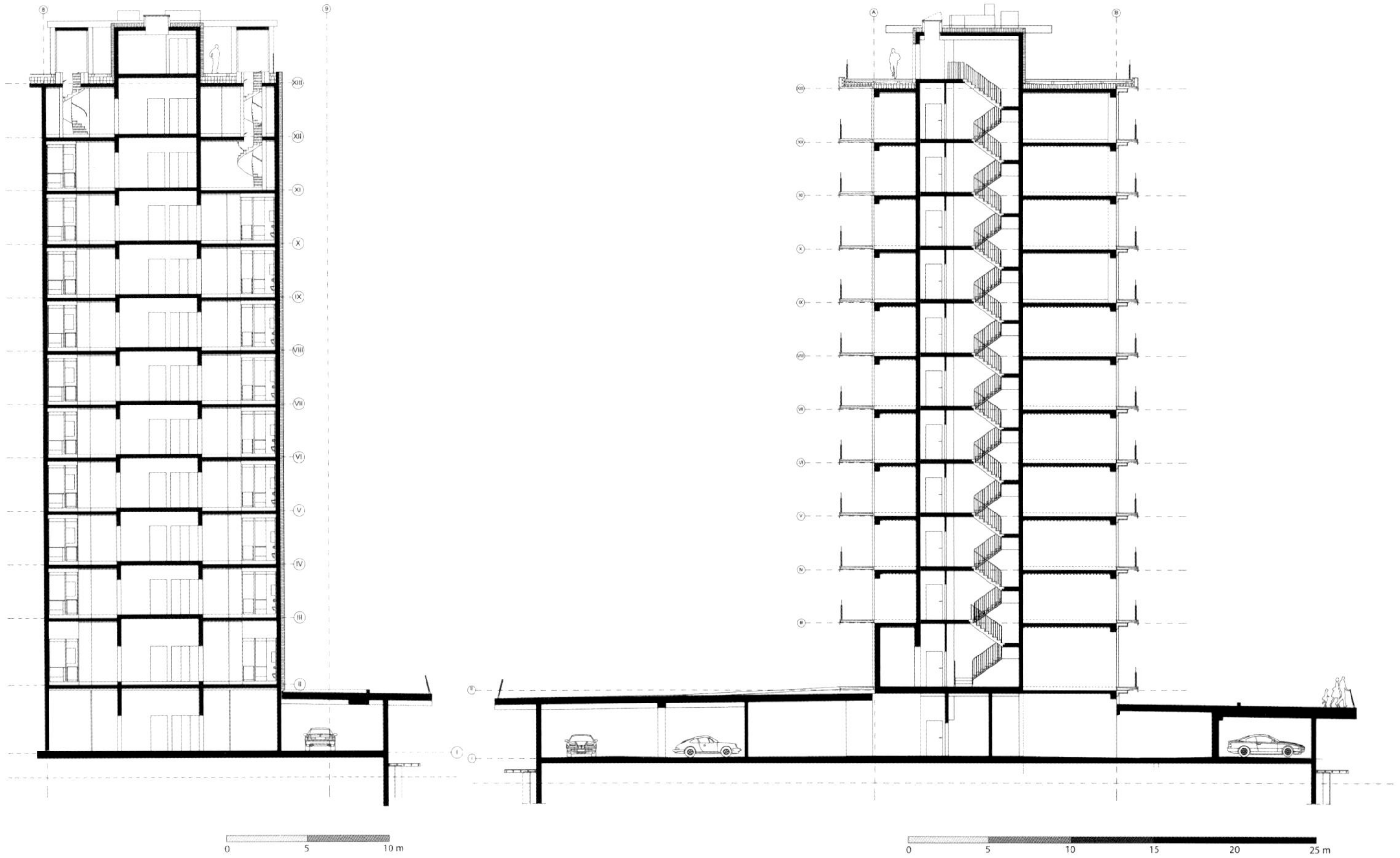

Section Plan 剖面图

Elevation Plan 立面图

depending on the weather. The shells are made from white cement concrete, marble dust and coloured glass fibre. In order to create a slender profile to this 11 storey high building, the plan of the apartments is conceived as 3 towers linked in a petal formation.

The building is oriented to take maximum advantage of daylight and is featuring large panoramic windows in the living rooms. The penthouse units have access to private roof terraces. Even the ground level units enjoy the spectacular views due to a 4-meter-high plinth, which conceals the parking garage. Surrounding most of the peninsula is a wooden bridge/marina allowing for temporary mooring. The facility is open to the public just as the peninsula itself. The project is meticulously thought through in order to keep the original design intend intact with the rational, pre-cast concrete element construction type. Because of the extensive use of repetitive modular components, "Metropolis" was built within budget.

哥本哈根城市公寓楼是个极具前景的城市住宅项目，坐落在哥本哈根港的一个名叫 Sluseholmen 的小半岛上，人们可以驾船直接到达城市公寓楼的正门。该地区在历史上曾是重要的工业区，如今已成为哥本哈根市中心最先进的住宅项目发展地。城市公寓楼位于阿姆斯特丹运河沿岸，坐拥美妙河景和怡人氛围。该地区显示出来的高档化还带来了无限商机，许多新商店、咖啡馆以及跨国公司都纷纷选择进驻这里。

丹麦建筑事务所 A/S 丹尼尔森建筑事务所和国际知名的未来系统事务所合作设计的这座城市公寓楼，高 11 层，是个纯住宅项目，共有 81 套公寓，大小从 90 m^2 到 210 m^2 不等，所有的单位都至少配有两个阳台。城市公寓楼的设计灵感来源于相依相伴的传统海港桥和楼群，以及周边的工业仓库。建筑采用曲线外形，纤维混凝土包层外壳由白色混凝土、大理石粉尘和玻璃纤维组成，会随着天气的变化呈现青蓝色。为了给这栋 11 层高的公寓打造出纤细的外形，设计师以花瓣造型把 3 栋大楼连接起来。

建筑的朝向设计力求最大限度地利用日光，并在客厅设置了大型的全景落地窗。空中别墅还享有私人的屋顶花园。另外，由于建筑的基座有 4 m 高，能把停车场很好地隐藏于其中，因此即使是置身低层的单位也可以尽享壮阔美景。半岛的四周围绕着一座滨海木桥和码头，便于临时锚泊。整个半岛和其设施都向公众开放。为了在保持原始设计完整的同时与合理的预制混凝土单元结构形式相匹配，该项目历经设计团队的大量的考察。由于重复模块化组件的广泛使用，城市公寓楼才得以在预期内竣工。

FEATURE 特点分析

Façade

The undulating blue façade of the practice's latest design echoes the sea and sky. The lines that the curves follow both vertically and horizontally across the façade are graceful and beautiful. There's something art deco about the layers of balconies and its porthole windows. The pigment is never distributed evenly. The panels have these variations in color. It is a common theme in Scandinavian craft and culture — to have a natural feel to materials.

立面

波浪起伏的蓝色立面呼应着港口区的大海与天空，优美而多变的曲线在水平和垂直两个方向上装点着立面，阳台和舷窗体现着其装饰艺术，玻璃混凝土覆面中的色料错落分散，与斯堪的纳维亚强调自然感觉的材料工艺文化相呼应。

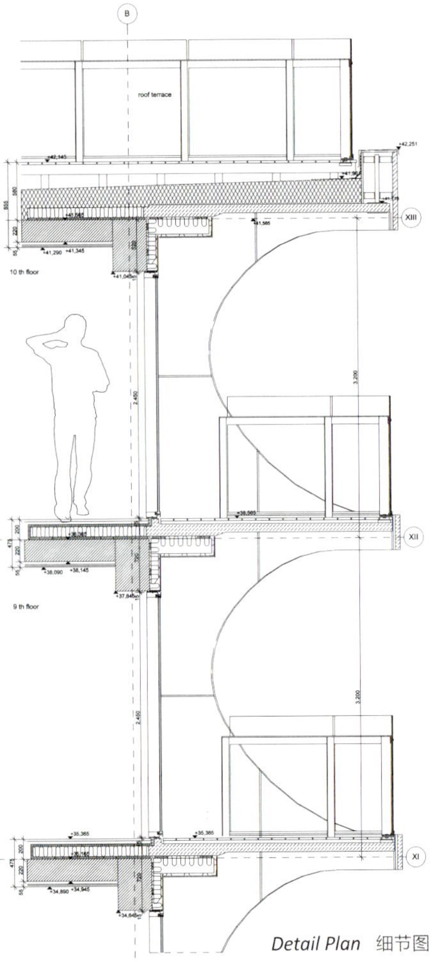

Detail Plan 细节图

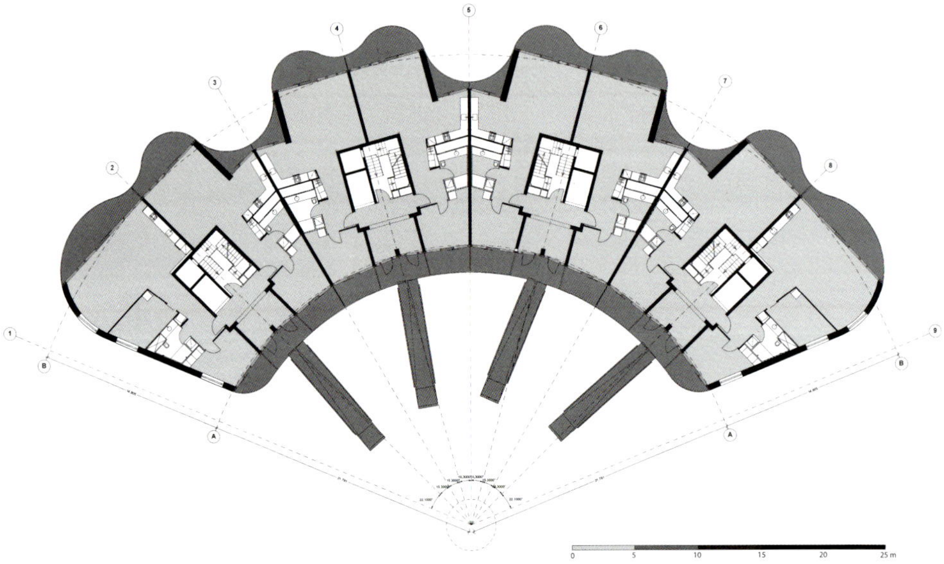

Ground Floor Plan 首层平面图

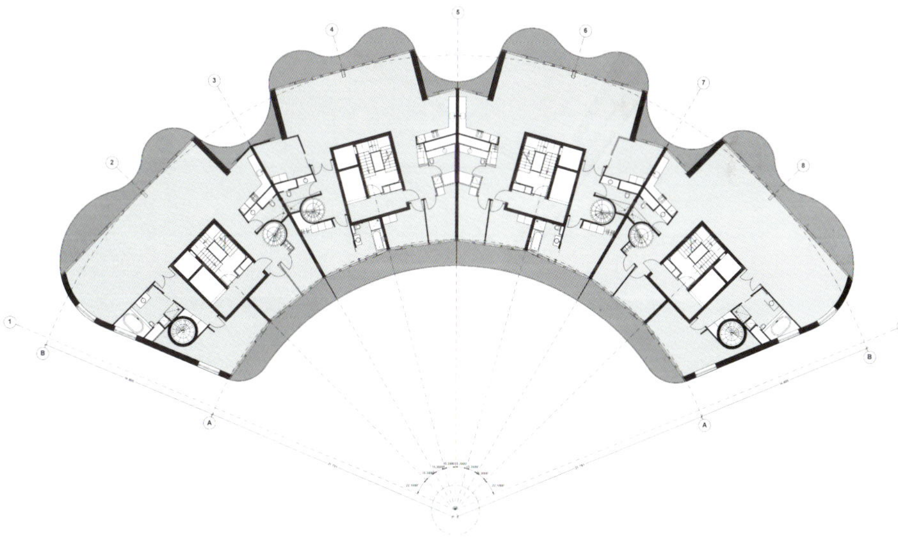

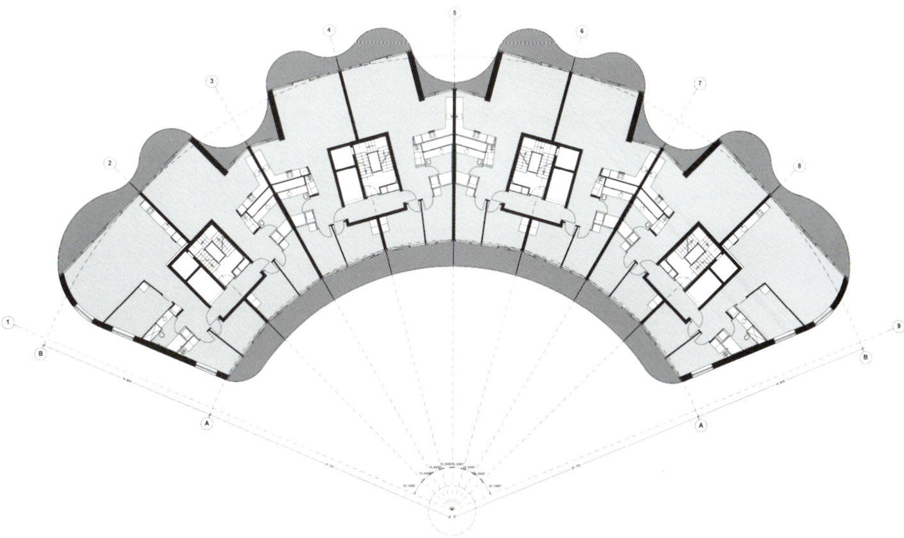

Living Foz Apartment Building, Porto, Portugal

Location: Porto, Portugal
Architect: dEMM arquitectura & Paulo Fernandes Silva
Gross Floor Area: 13,700 m^2
Number of Floors: 7
Photographer: Pedro Lobo

葡萄牙波尔图市福兹生活公寓楼

地点：葡萄牙波尔图
设计公司：dEMM 建筑事务所和保罗·费尔南斯德·席尔瓦建筑事务所
建筑面积：13 700 m^2
层数：7
摄影师：佩德罗·洛博

Surmount 突破点

BALCONY 阳台
FACADE 立面
MATERIAL 材料
ENERGY SAVING 节能

Living Foz is located in Foz do Douro, an Oporto city parish with ancestral occupation due to its south sun exposure and proximity to both the sea shore and the Douro River mouth. Nowadays, it struggles to consolidate and qualify its residential offer. Therefore, Living Foz was expected to present a new approach to high quality residential buildings, emphasizing the relationship between interior and exterior, taking advantage of the sea shore views and the Portuguese sunny climate.

Accomplishing these expectations, the balcony angle articulation creates outdoor spaces that are enriched by contrasts of light and shade, exposure and protection.

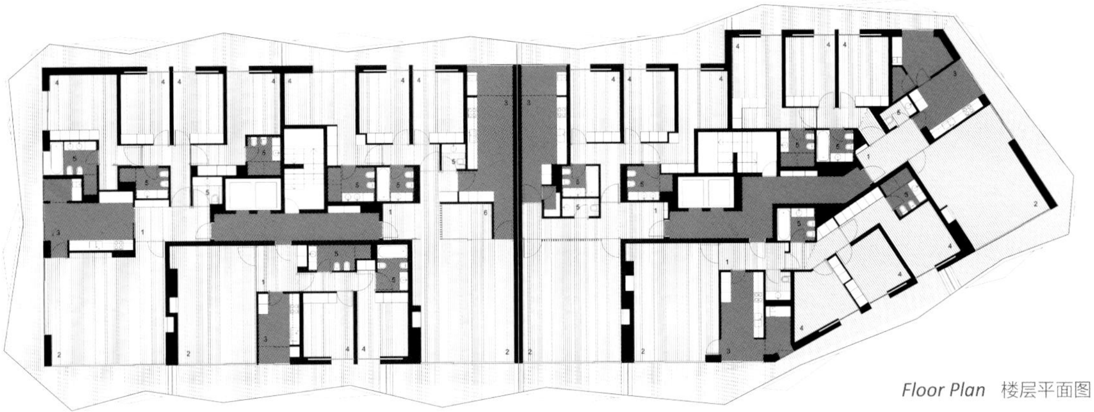

Floor Plan 楼层平面图

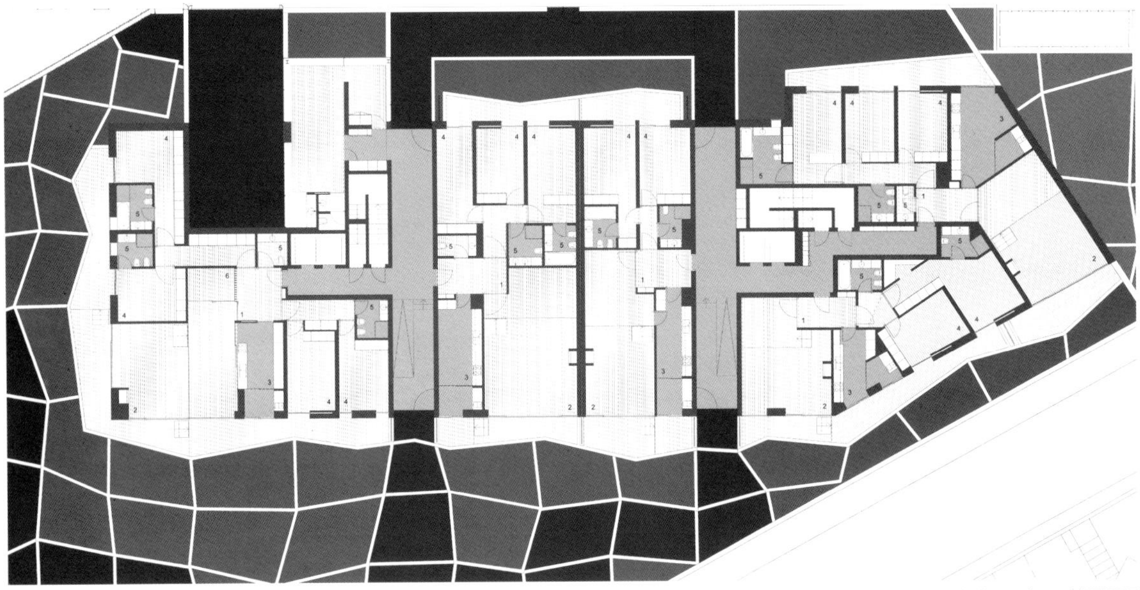

Ground Floor Plan 首层平面图

The interior compartments, articulated in order to encourage a flexible use of the social areas, are widely opened to the exterior by large floor to ceiling sliding windows. Nevertheless, these glass surfaces are always protected from direct sun exposure, representing a simple though effective way to reduce energy consumption.

Unleashing 30% of the plot area for a collective use garden, a relief to the neighboring construction is obtained, qualifying the surrounding residential area. The geometric and material principle of the building continues through the external areas, defining the direction and shape of the garden paths.

The 40 apartments (divided in typologies of 2 to 5 bedroom) are distributed over 7 floors and completed with private parking in 3 underground floors. The pavement is lined in dark varnished wood, and the walls and carpentry are painted in a very light white tone. Kitchens and toilets are coated with Portuguese natural stone.

Living Foz is also about sustainability, regarding the use of high technology insulation in windows, walls and covertures as well as the conception of the building design itself. In fact, the controlled sun exposure, the thermal inertia of the façade, the use of alternative energy and electrically efficient appliances, are features that integrate the building in a quality standard of respect for the individual and their habitat.

福兹生活公寓楼坐落在波尔图市教区的福兹－杜多罗市，其南部阳光充足，邻近海边和多罗河口，当地人世代生活在这里。如今，这座城市的住宅条件需要得到巩固和优化。因此，福兹生活公寓楼将会加强室内和室外的联系，充分利用海景优势和葡萄牙阳光充足的气候，以新的方式诠释高品质公寓楼。

为了达到此目的，设计师使阳台以一定角度连接，打造出光影分明的室外空间，既能采光又能遮阳。

考虑到社交空间的灵活运用，室内空间被连接起来，使用可滑动的大型落地窗连通到室外。这些玻璃表面能防止太阳直射，简单而有效地减少能源消耗。

此外，30% 的小区面积被腾出来用来作公共花园，以减轻周围建筑的负担，同时也提高周围的住宅区的品质。沿用于建筑外部区域的几何图样和材料，决定了花园小径的铺设方向和形状。

40 套公寓（按照种类分成 2~5 个卧室）分布在 7 层楼，地下 3 层则是私人停车场。人行道用的是暗光漆木，墙壁和木器则选用淡白色基调。厨房和厕所用的是葡萄牙天然石材。

福兹生活公寓楼也同样注重可持续发展，就像建筑本身的设计理念一样，窗户、墙壁和覆盖层用的都是高科技绝缘材料。事实上，控制太阳直射强度、立面的热惯量，使用可替代能源和节电的家电都能使建筑表现出对这里的居民和他们的家园的高度尊重。

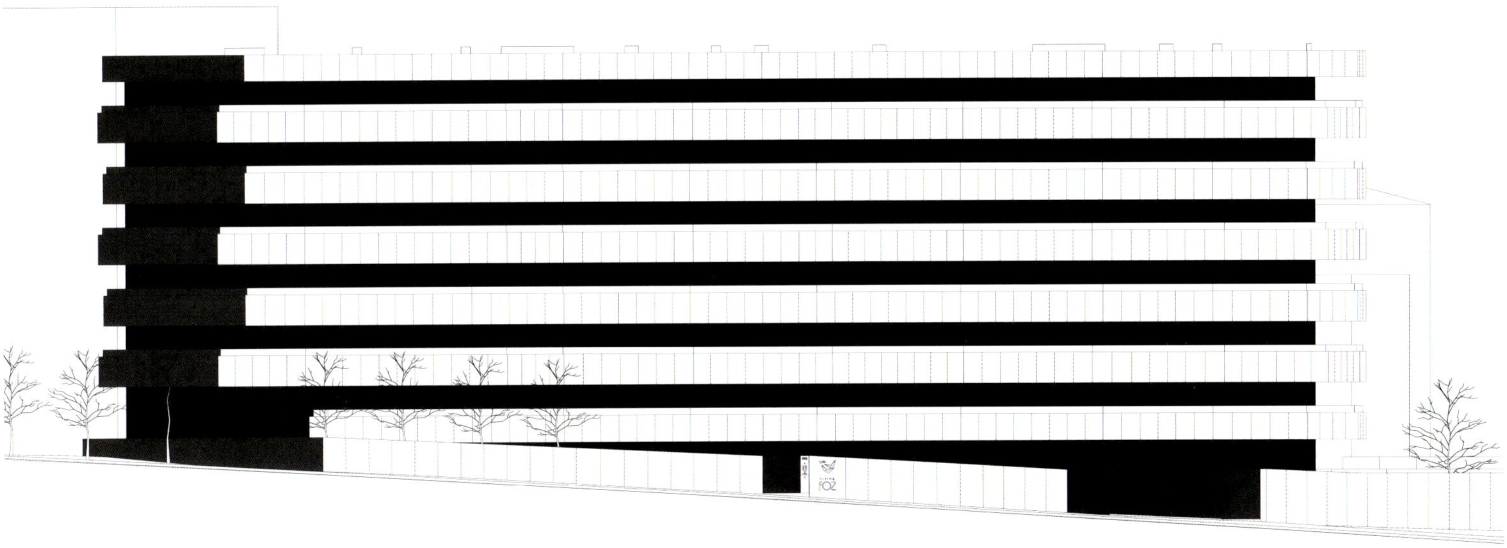

Elevation Plan　　立面图

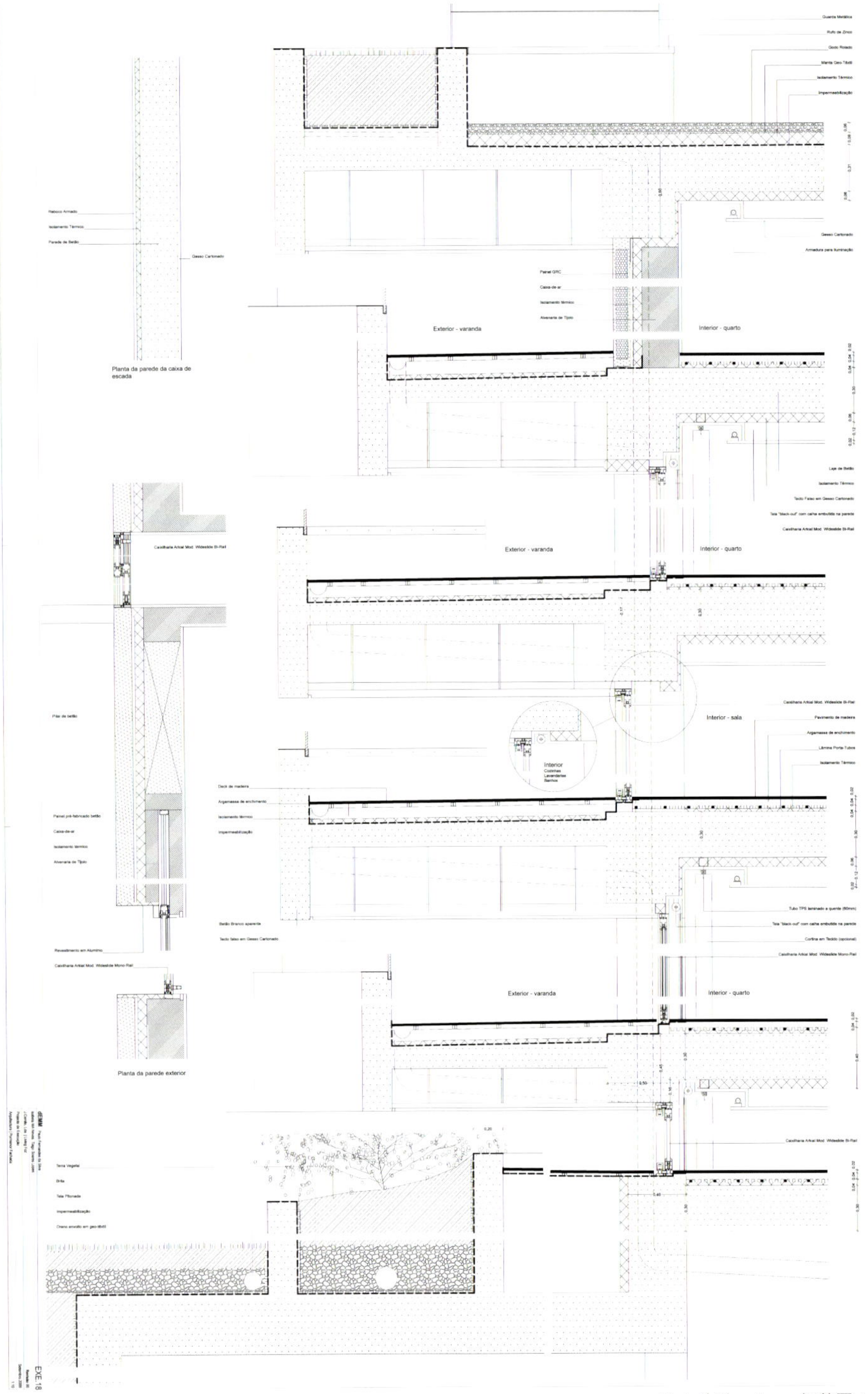

Detail Plan 1　细节图 1

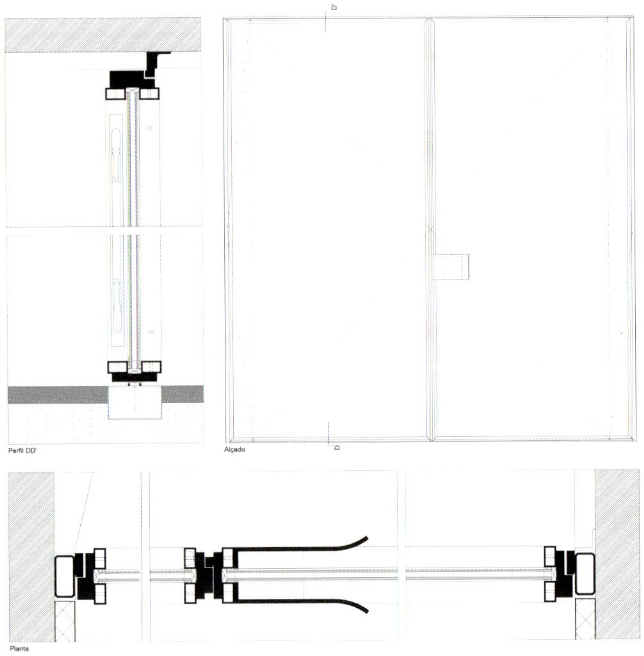

Detail Plan 2　细节图 2

FEATURE 特点分析

Material

There is a contrast between the white cast-in-place concrete and the dark Glass Reinforced Concrete panels. The shape of the balconies is emphasized. The species were chosen to create a wide variety of colors and textures, and to expose or dissimulate the public and private spaces of the ground floor. The finishing materials are intended to be durable and to reproduce in the interior the contrasts between light and shadow of the exterior.

材料

白色现浇混凝土和钢筋混凝土镶板上的深色玻璃形成鲜明对比，强化了阳台的外形；不同材料的选择创造了不同的颜色和纹理，展示了首层公共场所，并隐藏起私人空间；饰面则选用了耐用，且能在室内重现外部光影对比的材料。

Social Housing 100, Paris, France

Location: Paris 19ème arrondissement
Architect: Jakob + MacFarlane Architects
Gross Floor Area: 6,600 m²
Number of Floors: 6
Photographer: Nicolas Borel & James Ewing

法国巴黎 100 社会住房

地点：巴黎第 19 区
设计公司：雅各布和麦克法兰建筑事务所
建筑面积：6 600 m²
层数：6
摄影师：尼古拉斯 · 波雷尔、詹姆斯 · 伊文

Surmount 突破点

STRUCTURE 结构

FACADE 立面

ECOLOGY 生态

MATERIAL 材料

In the context of the reconversion of the ancient site of the hospital Hérold, conceived by Philippe Madec in Paris 19th arrondissement, Jakob & MacFarlane establish/design three mineral forms standing free on a kind of slope triangular island planted with some trees. These three forms, grow like roots on portions of the territory that have been freed by the constraints of the site: protected trees, the ancient existing wall of the beltway, the size and prospect, the not buildable surfaces or the difference in height. Their shapes are at the same time the result of these constraints and a matrix structure generated by an irregular grid on the floor.

The organization of the vertical orthogonal plans is the result of this grid and of the limitation of the territory; it forms the carrying structure of the buildings. The structure develops at the same time a thick base and horizontal plans that are formed by irregular segments on a regular frame that result in the levels of the floors.

Three blocks up to 5 and 6 floors high result out of this rough skeleton in concrete, in opposition to a dense environment. From an urban point of view, these autonomous rocks favor the porosities towards the heart of the isle and recreate a discontinuity against the wall.

The silhouette and façade of the buildings are delimited by faceted outlines. Starting from the 4th floor the profiles get thicker and thinner. The outside walls develop similar too long segments dissociated from the floor by a hollow joint. The best oriented façades are set up by multiple bays off all heights.

The apartments are protected from the urban space of the Boulevard Serurier by different vertical filters: the outer wall, the trees, the railing that is slightly set back from the edge of the balcony, the ETFE curtain that creates a winter garden between the deep balcony, treated as a gangway and the façade.

There are a number of contractions that result into a terrace on the last floor. Combined with aluminum carpentry interrupted by thermal bridges and an exterior isolation, those contractions make it possible to improve the thermal performances of the envelope, also by the intrinsic organization of the utilities. It is the same for the roofs made of solar panels and planted with sedum. The passive role of the ETFE, used as a curtain on a fixed rail, protects the balconies from the wind. The outside spaces become at the same time a sort of micro ecosystems and technical support in order to recuperate the rain water. The edges of the balconies are made of porous volcanic stone and musk, a watering system is also integrated. The garden, situated on the ground-floor and creating a huge green platform in the heart of the isle, follows the same principle of ecosystem. A walkway goes along the façades that are oriented towards the boulevard, perforates the isles, surrounds and joins them. The ground shapes topography of a terrace garden, delimited by a low mineral strip and protected by a supporting wall in millstone. There are multiple green spaces, arranged according to a grid from which come out high stems and trees.

The bed and living rooms are generally oriented towards the heart of the isle. The apartments develop from the vertical circulations in the heart of the buildings that lead into seven apartments.

From a typological point of view, there is certain fluidity in the plans. Sliding walls define the space and liberate the façades into a plan of continuity. They allow more permeability and flexibility for the user. The corridors are reduced to the minimum. The entry is a point of distribution opening on the big diagonals of the living room and the other rooms. Hence, the attention is drawn towards the notion of panorama. This will to make the space bigger is also present on the small surfaces. As soon as one enters, the look is drawn towards the outside. Finally, the project is unified by the "earth" color of the railing, outside walls and façade panels.

在复原菲利普·马代克设计的巴黎第 19 区埃罗尔德医院的背景下，雅各布和麦克法兰建筑事务所设计了三栋矿物外形的建筑，随意伫立在斜坡三角岛上，周边种植了树木，这三栋矿物外形的建筑仿佛扎根于地面，并从地面的限制中解放出来一样。这个地方的制约因素包括被保护的树木、环道的现存古墙、建筑的大小与景色、不可采用的表皮以及高度差。它们的外形既是这些限制因素的结果，又是由地板上的不规则网格构成的矩阵结构。

由于网格和场地的限制，在垂直方向上采用了直角结构，形成建筑的支撑结构。这种结构发展成了厚实的基座以及在规则框架下的不规则片段形成的平面，构成了楼层。

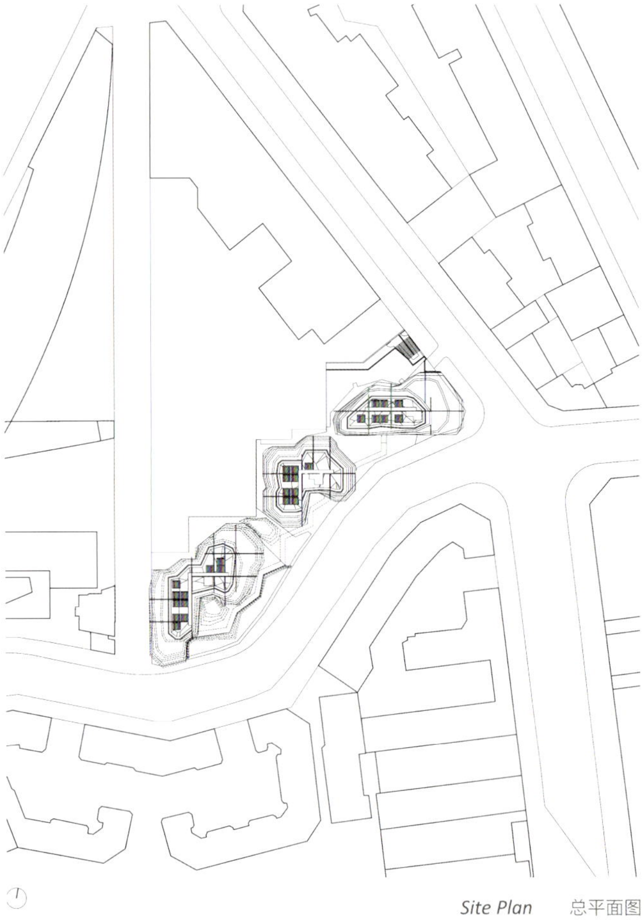

Site Plan 总平面图

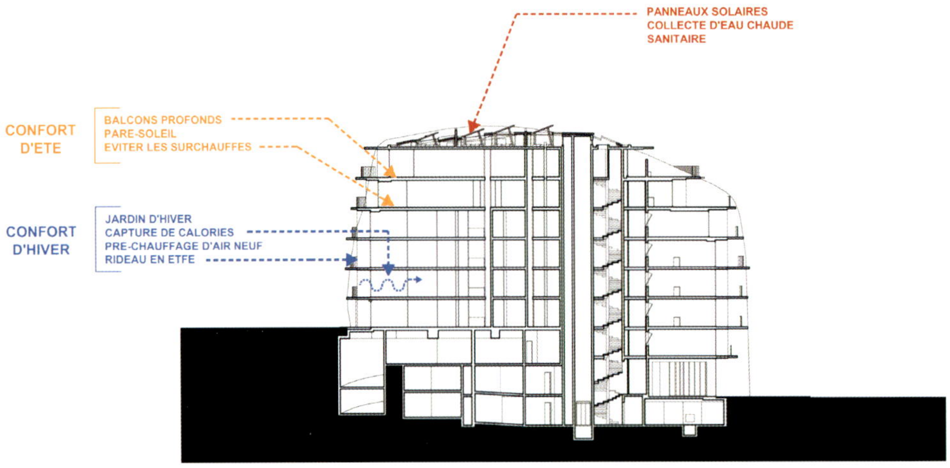

Section Plan 1 剖面图 1

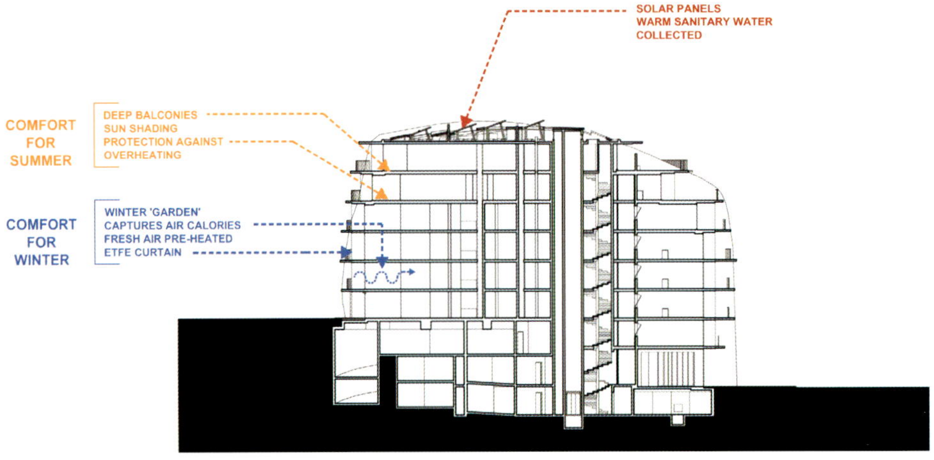

Section Plan 2 剖面图 2

三栋楼都是粗混凝土结构，高 5~6 层，与周边的密集环境形成对比。从城市角度看，这些独立的岩石在岛中心塑造了孔隙，并重现了外墙的不连续性。

建筑物的轮廓和立面被多面分割，从 4 楼开始外形发生改变，有的地方变厚，有的地方变薄。外墙也呈长形，并被地板和外墙之间的空隙所隔开。朝向最好的立面有多个间隔。

为了保护公寓，使其不受苏洛伊大道的喧闹所影响，设计师使用了各种垂直过滤方式：外墙、树木、从阳台边上后缩的栏杆和乙烯幕墙。乙烯幕墙既是一种立面，又使阳台变成过道，在较深的阳台里创造了一个冬季花园。

因为有许多缩进，所以公寓顶层安排了一个露台。使用被热桥和外部隔离装置阻断的铝木，利用这些缩进和公共设施的内部组织提高了建筑表皮的热效应。种有植物的太阳能面板屋顶也是如此。固定轨道上的乙烯窗帘，则起到御风的作用。外部空间既成为微生态系统，也成为一种技术支持，确保了雨水的回收。阳台的边缘由多孔的火山石和香木堆砌，还整合了雨水回收系统。首层的花园产生了一个位于岛中心的巨大绿色平台，遵循着同样的生态系统原则。一条沿着立面朝向大马路的走道，穿过岛屿，围绕着它们并把它们结合在一起。首层宛如一个由低矿物带划分的梯台花园，被磨石支撑墙保护着。此外还有许多根据网格安排的绿地。现有的树和矗立着的建筑都在这个网格规划中。

卧室和客厅都面向岛中心。公寓楼根据建筑中心进行垂直环绕设计，分为七个公寓。

从类型学的观点来看，这个设计有一定的流动性。滑动玻璃墙划分了区域空间，并使立面连续一致，让住宅更加通透、灵活。而走廊则设计到最小，入口开在客厅和其他房间的对角线上，因此，人们的注意力就会放在室内的全景上，使空间在小小的外观上显得更为宽敞。一进入里面，视线就会被自然地引向外面。最后，该项目统一使用了“大地色”的栏杆、外墙和立面面板。

FEATURE 特点分析

Structure

Grass and trees are planted in the grid to meet the limits of the project. The grid units the buildings on the three irregular bases.

结构

绿地、树和建筑都处在网格中，以适应该项目的限制要求，并把坐落在三片不规则基地上的建筑统一起来。

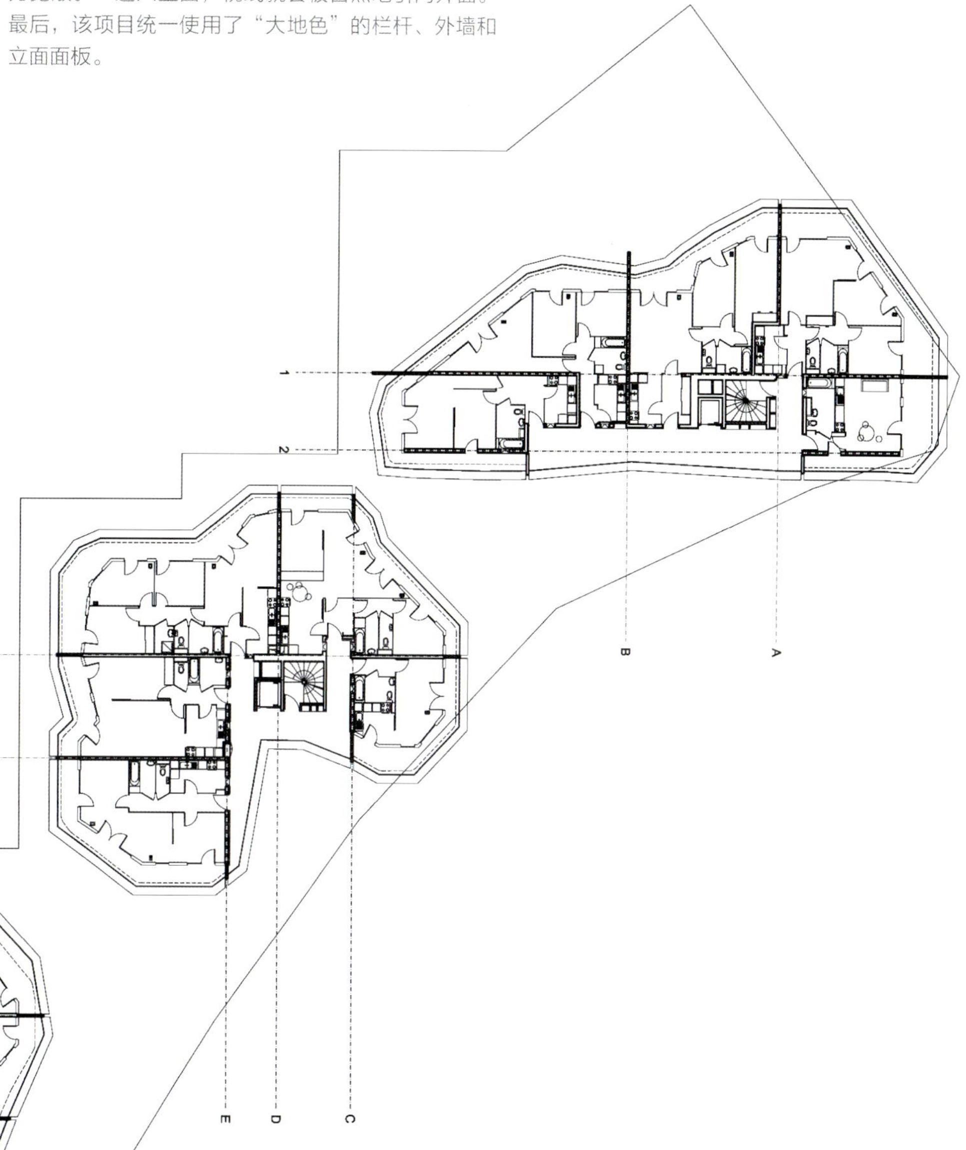

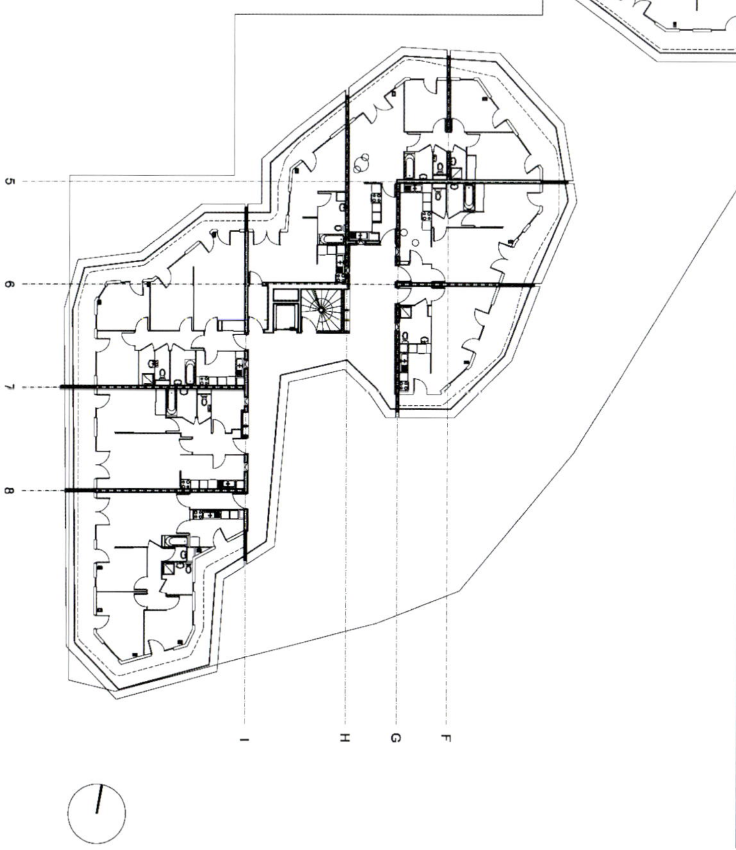

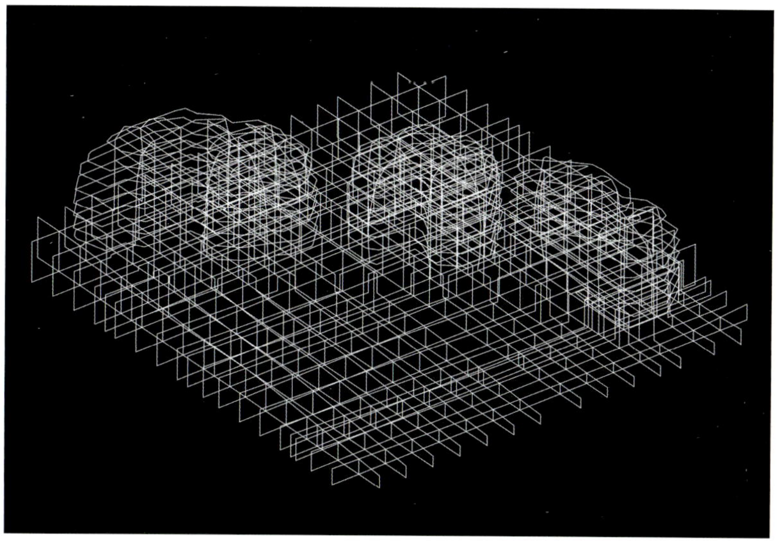

Concept Plan 概念图

Eliza Apartments, Sydney, Australia

Location: Elizabeth Street, Sydney, Australia
Architect: Tony Owen Partners Architects
Number of Floors: 17

澳大利亚悉尼市伊丽莎公寓

地点：悉尼伊丽莎白大街
设计公司：托尼·欧文与合伙人建筑事务所
层数：17

Surmount 突破点

LUXURY 奢华

FACADE 立面

BALCONY 阳台

SKY VILLA 空中别墅

The Eliza luxury apartments by Tony Owen Partners have just been approved in Sydney. The Eliza is located on Elizabeth Street in Sydney's CBD on prestigious Hyde Park.

With only 19 apartments, this exclusive 17-storey development will create a new standard in luxury and contemporary design for Australia. The units have a unique parametrically designed façade. It responds to the changing solar, view and planning conditions on each level so that the façade on every level is different. The design features 2 and 4 bedroom units with panoramic park and harbour views as well as a 3 storey penthouse.

The penthouse has been conceived as a unique world class address with roof top pool, private elevator and grand staircase and ceiling heights. For design inspiration the designers looked to the classics of 20th century classic design to create a world class building for a truly unique site. This project is one of a half dozen towers by Tony Owen Partners in the Sydney CBD which feature digital design practices to create unique and exciting environmental solutions in a new approach to urban design. Construction is now commencing on site and is expected to be completed by 2012.

The Eliza Apartments is one of the most elegant and prestigious apartment buildings in Sydney. The project combines classic 20th century style with the latest parametric design tools to create an iconic landmark. The site is located on Sydney's historic Elizabeth Street on Hyde Park. This is Sydneys most exclusive address. This design at once responds to the historic streetscape, and massing considerations of the site, at the same time the design relates to the unique lifestyle and character of the surrounding area.

The profile of every level is different. This responds to the differing unit plans and views on each level. As the views change progressively up the building, so does the shape. The 17-storey building contains 19 units with most levels having one unit per floor and a 3-storey penthouse on top. The tapering shape of the penthouse maximizes solar access to Hyde Park and terminates the fluid geometry.

The area includes a number of heritage structures representing several periods, some flamboyant in style. Whilst not attempting to mimic its neighbours, the design presents a contemporary expression of the principles embodied in the surrounding buildings. The design uses traditional and elegant contextual materials, yet in a contemporary and challenging way.

The lower 4 levels are expressed as a solid sandstone podium to respond to the solid expression of the Sheraton and Great Synagogue to the south and adjoining buildings to the north.

The result is a thoroughly contemporary parametric design that creates a timeless design in a classic Sydney area.

由托尼·欧文与合伙人建筑事务所设计的伊丽莎豪华公寓位于著名的海德公园中央商务区的伊丽莎白大街，已在悉尼被批准。

尽管这栋独特的 17 层高的建筑仅拥有 19 套公寓，但它将为澳大利亚现代奢华的风格设计创立新标准。这些公寓有一个独特的参数化设计立面。根据太阳位置的变化，每层的景观和设计条件决定了每层的立面都不尽相同。这 19 套公寓里包括了能欣赏到公园和海港全景的双卧和四卧公寓，以及一个三层高的空中别墅。

空中别墅的设计独特而一流，带有空中泳池、私人电梯、豪华楼梯以及高高的天花板。在设计方面，设计师从 20 世纪的经典设计中汲取灵感，创造出一栋世界一流的建筑，以体现这里独特的地理位置。在悉尼中央商务区，托尼· 欧文与合伙人建筑事务所使用数字设计的新方法，因地制宜，不断创造可行的环境解决方案，而这栋建筑就是其中的一个项目。建筑已经开始施工，预计 2012 年完成。

伊丽莎公寓是悉尼最优雅、最著名的公寓楼之一。结合 20 世纪的经典风格和最新的参数化设计工具，该项目成为了悉尼中央商务区的一栋标志性建筑。该项目位于悉尼海德公园历史悠久的伊丽莎白街，是悉尼最高级的地方，项目的设计与历史悠久的街景相呼应，综合考虑了整个区域的特点，同时体现了它与周边地区独特的生活品质与生活方式的完美结合 。

由于公寓中每层的单元设计和视野都大不相同，因此每一层的剖面自然也都有所差异。视野会随着层数的不同而逐步发生变化，外形也是如此。这栋 17 层的建筑中，多数楼层只有一套房，而最上方则是三层高的空中别墅。

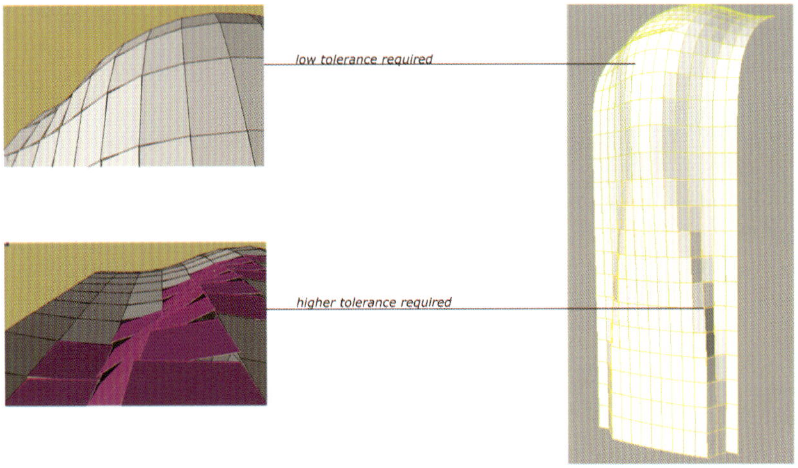

Façade Analysis 1　立面分析图 1

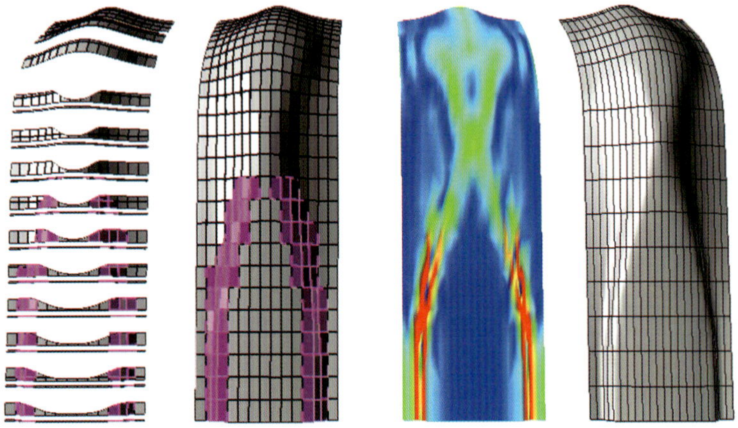

Façade Analysis 2　立面分析图 2

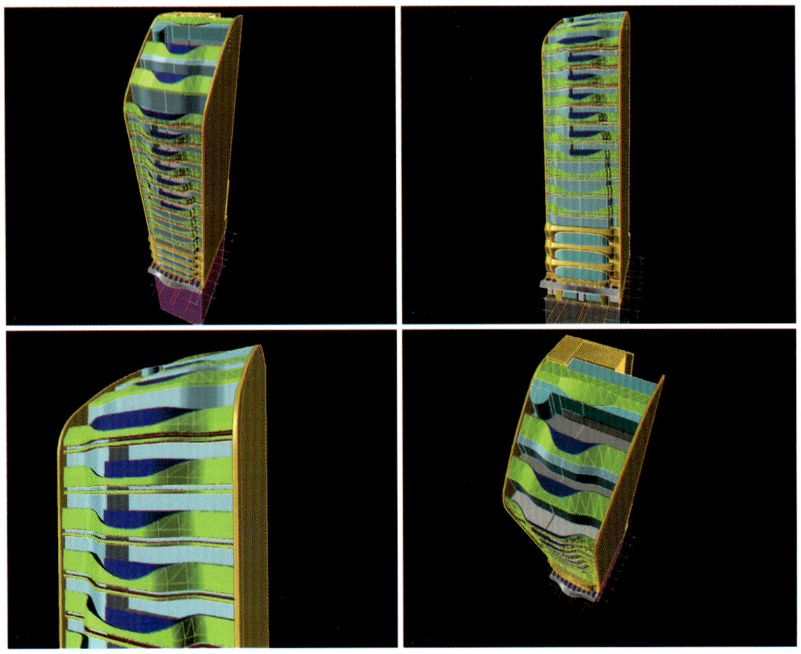

Façade Analysis 3　立面分析图 3

Calibre
ST. JAMES HOTEL
RECESSION BUSTER
$9.95
Meals
Friday Nights
$4 Drinks
$9 Cocktails
Chesterton
FOR SALE
PRIME CITY FREEHOLD
8238 0000

空中别墅向上逐渐细化的形状，使海德公园尽可能多地沐浴在阳光之中，并使流动的几何形体在此终止。

该地区有许多遗迹，代表几个不同的时代，其中不乏一些风格华丽的建筑，但此项目的设计却没有试图模仿它们，而是以一种现代化的方式体现了它与周边建筑一致的设计原则。总体而言，整个公寓的设计采用了传统而优雅的与背景一致的材料，极富现代性和挑战性。

下面的四层裙楼用了实心砂岩，以搭配南面气势宏伟的喜来登酒店和犹太大教堂，同时与北面的毗邻建筑相呼应。

这栋完全的当代参数化建筑，最终在悉尼经典地区创造了超越时间界限的设计风尚。

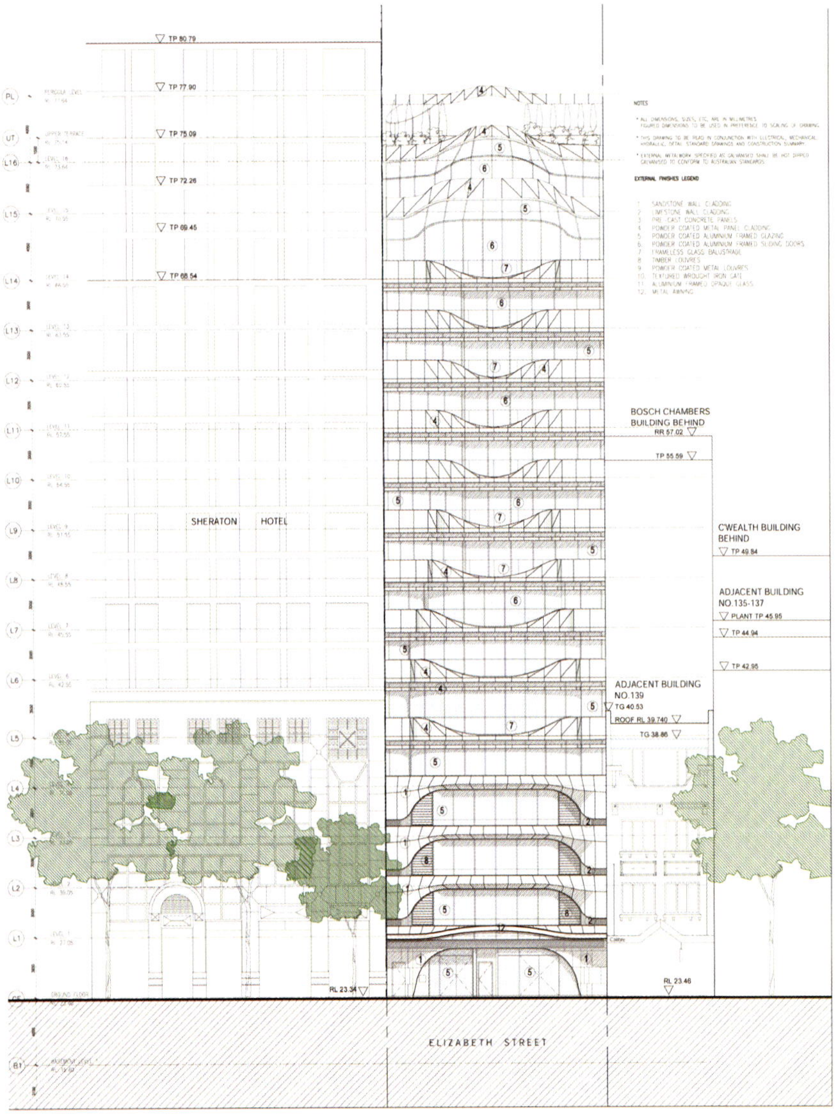

East Elevation Plan 东立面图

FEATURE 特点分析

Façade

3-D modelling software and scripting is used to create a façade that is progressively responsive to changing conditions. The clean lines of the façades and subtle curves create a bold and sophisticated statement, which is a contemporary expression of the traditional bay window.

立面

设计师使用三维建模软件和脚本，创建了随着环境变化而变化的立面，干净的立面线条和微妙的曲线，生动而高雅，是传统的凸窗的新式表达。

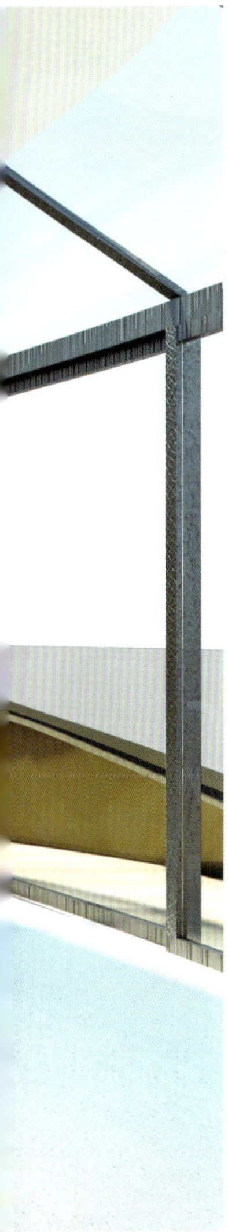

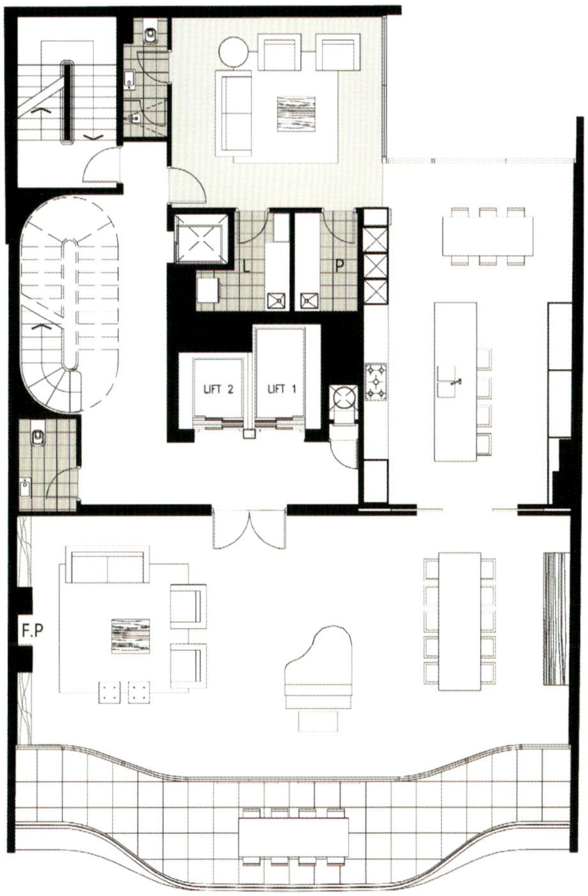

14th Floor Plan 14 层平面图

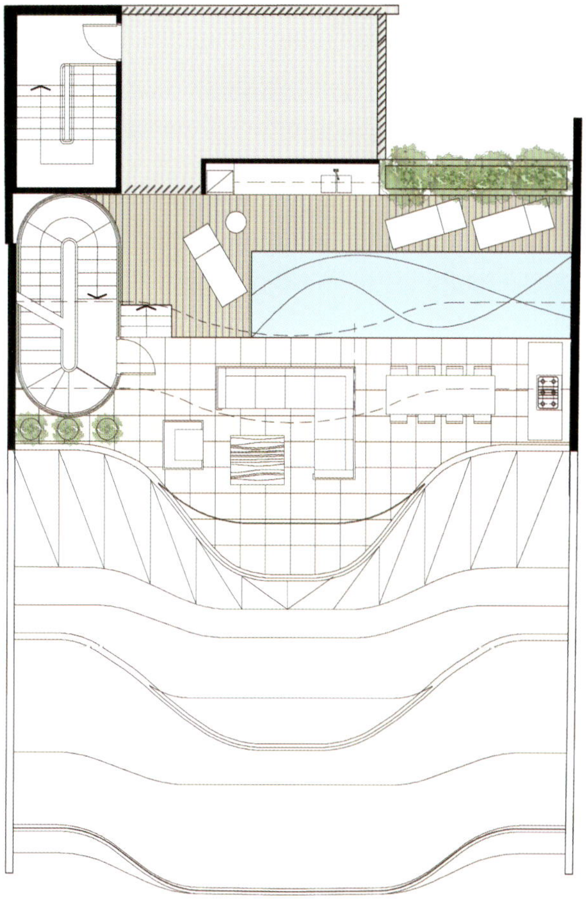

16th Floor Plan 16 层平面图

1st Floor Plan 1 层平面图

8th Floor Plan 8 层平面图

Apartmenthouse Joh 3, Berlin, Germany

Location: Johannisstraße in Mitte ,Berlin
Architect: J. MAYER H. Architects
Number of Floors: 7
Photographer: Ludger Paffrath

德国柏林 Joh 3 新公寓

地点：柏林 Mitte 市中心区 Johannisstraße 大街
设计公司：J · 迈耶 · H 建筑师事务所
层数：7
摄影师：路德格尔 · 帕夫若斯

Surmount 突破点

SHAPE 造型

SKIN 表皮

BALCONY 阳台

MATERIAL 材料

Property development group Euroboden is building a unique apartment house at Johannisstraße in Mitte, Berlin's downtown district. J. MAYER H. Architects' design for the building, which will soon neighbor both Museum Island and Friedrichstrasse, reinterprets the classic Berliner Wohnhaus with its multi-unit structure and green interior courtyard. Plans for the ground floor facing the street also include a number of commercial spaces. The generously sized apartments will face south-west, opening themselves to a view of the calm, carefully designed courtyard garden. Spacious, breezy transitions to the outside create an open residential experience in the middle of the city that, thanks to the variable heights of the different building levels, also offer an interesting succession of rooms. The units' varying floorplans and layouts indicate a number of housing options; condominiums are organized into townhouses with private gardens, classic apartments or penthouses with a spectacular view of the old Friedrichstadt. The integrated design concept, which incorporates everything from façade to stairwells, elevators to apartment interiors, promises a unique spatial and living experience with an eye to high design.

Euroboden 物业公司正在柏林市中心 Mitte 区的 Johannisstraße 大街建造一栋独特的公寓楼，附近是博物馆岛和弗里德里希大街，由 J · 迈耶 · H 建筑设计公司担任设计，重新诠释了经典柏林旧式的多单元结构和室内绿色庭院。临街的首层布局包含了商业空间。宽敞的公寓面向西南方，朝向精心设计的庭院，享受一丝的宁静。由于周边建筑的高度不同，宽敞的过道不时吹来微风，给建筑创造了一种位于市中心的开放住宅体验，还提供给房间有趣的连续性布局。不同的户型和布局提供了多种选择，有带私家花园的别墅、经典型公寓和能欣赏到老弗里德里希街区的空中别墅。综合的设计理念把立面、楼梯井、电梯和公寓内饰融合在一起，在打造了高端设计的视觉盛宴的同时，呈现出一种独特的空间和生活体验。

FEATURE 特点分析

Skin

The sculptural design of the suspended slat facade draws on the notion of landscape in the city, a quality visible in the graduated courtyard garden and the building's silhouette and layout.

表皮

悬板式立面雕塑的设计借鉴了城市景观的概念，在渐变的庭院和建筑的轮廓、布局上反映了高质量的设计。

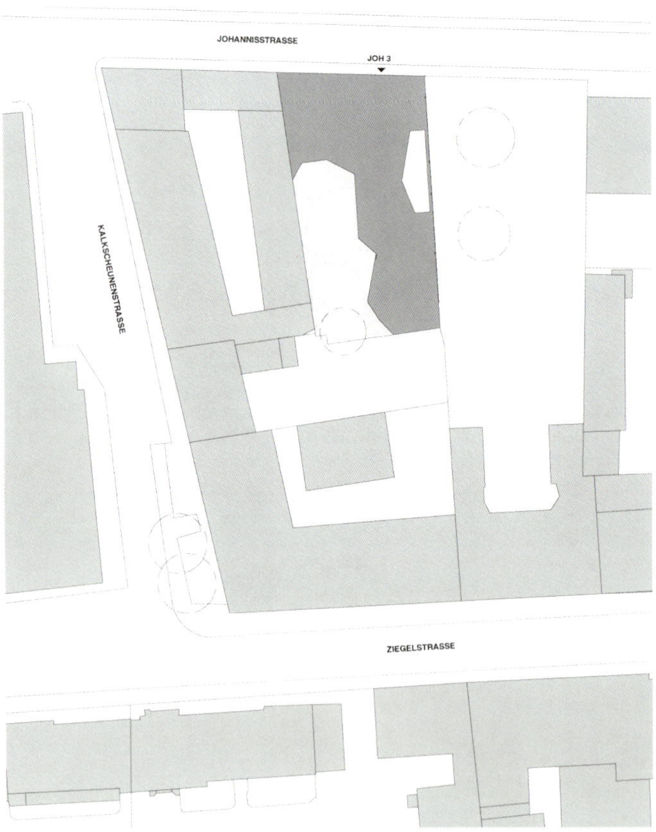

Site Plan 总平面图

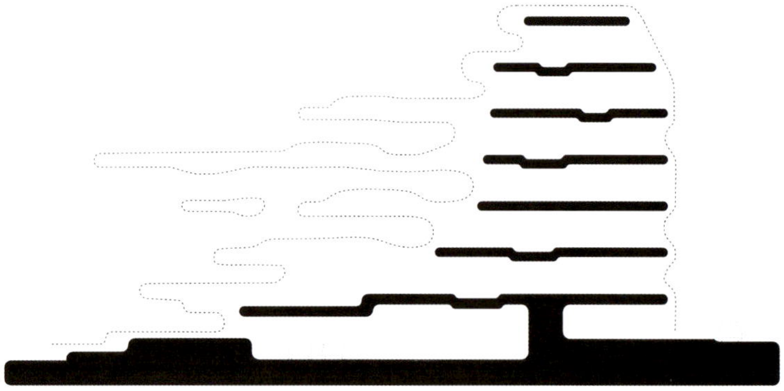

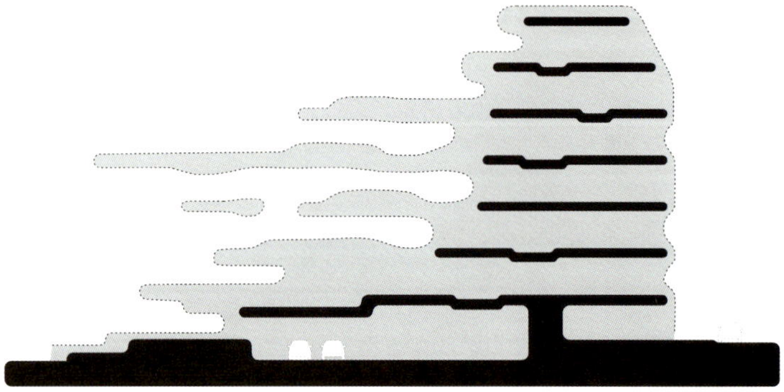

Analysis Plan 分析图

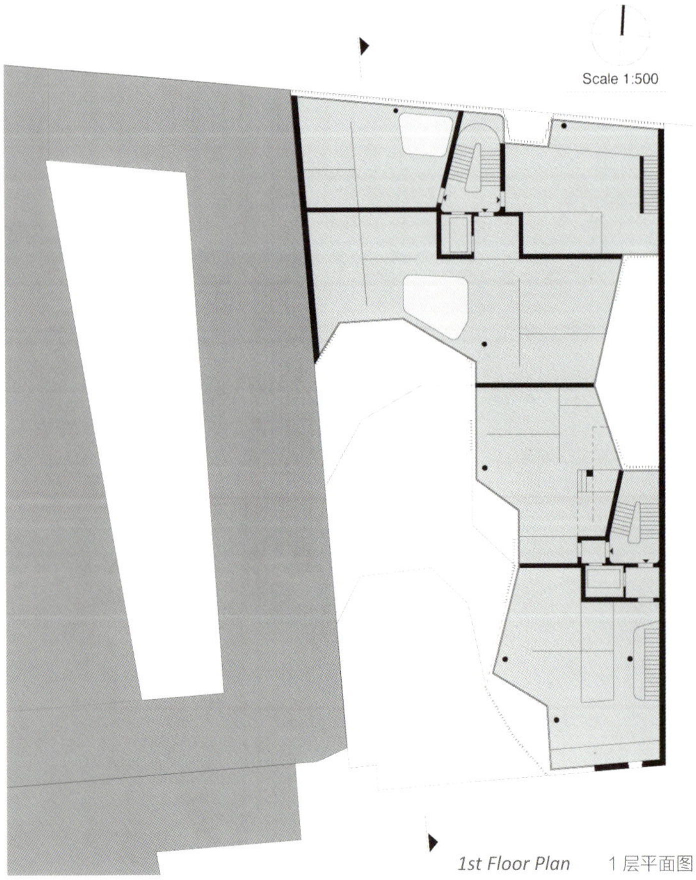

1st Floor Plan 1 层平面图

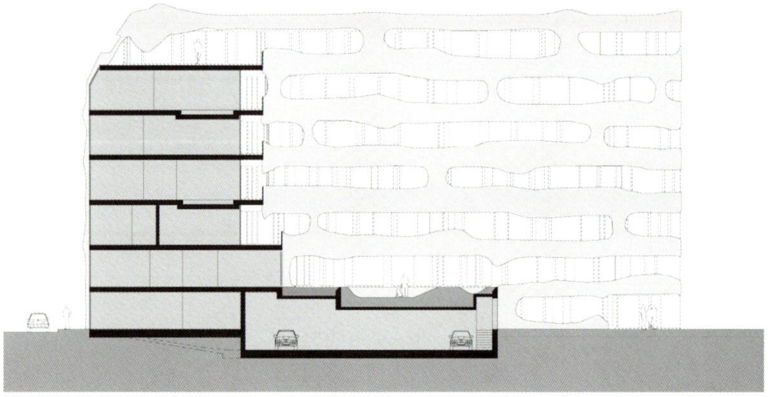

Longitudinal Section 纵剖面图

Elevation Plan 立面图

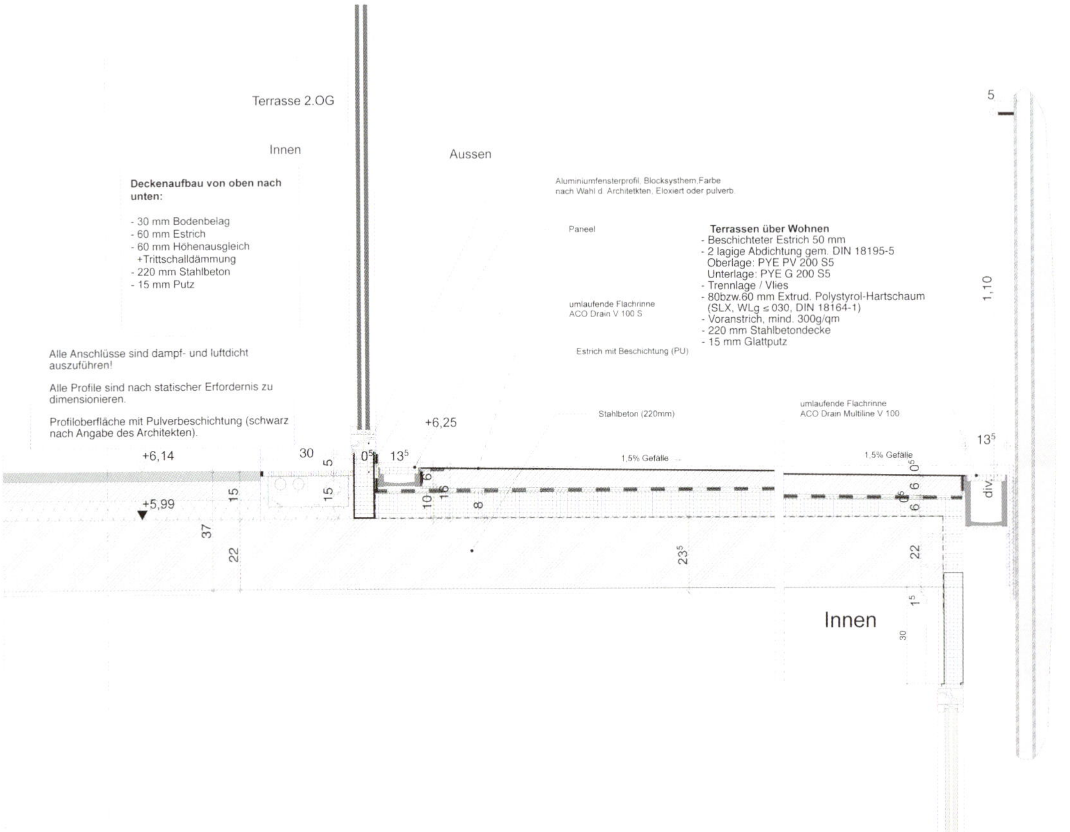

Detail Plan 细节图

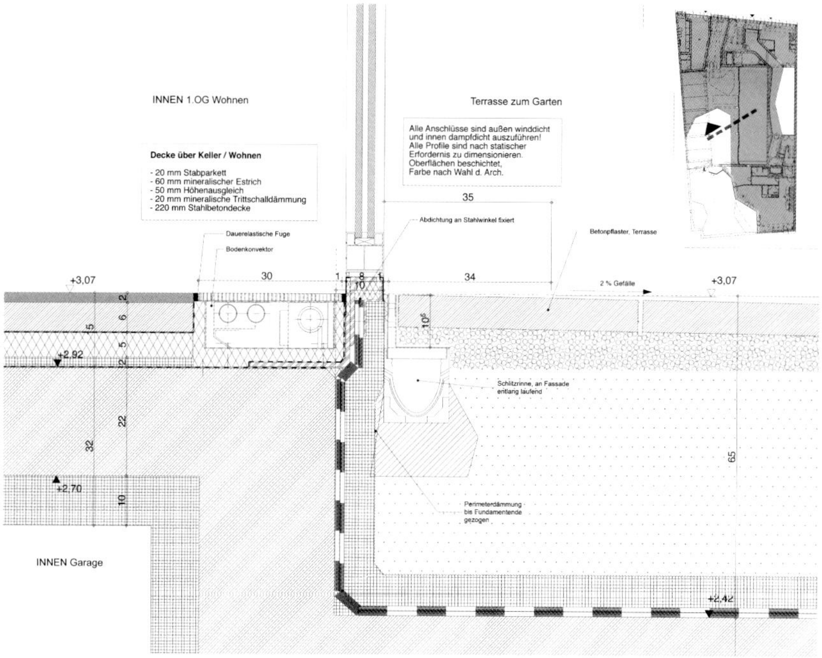

Detail 1 细节图 1

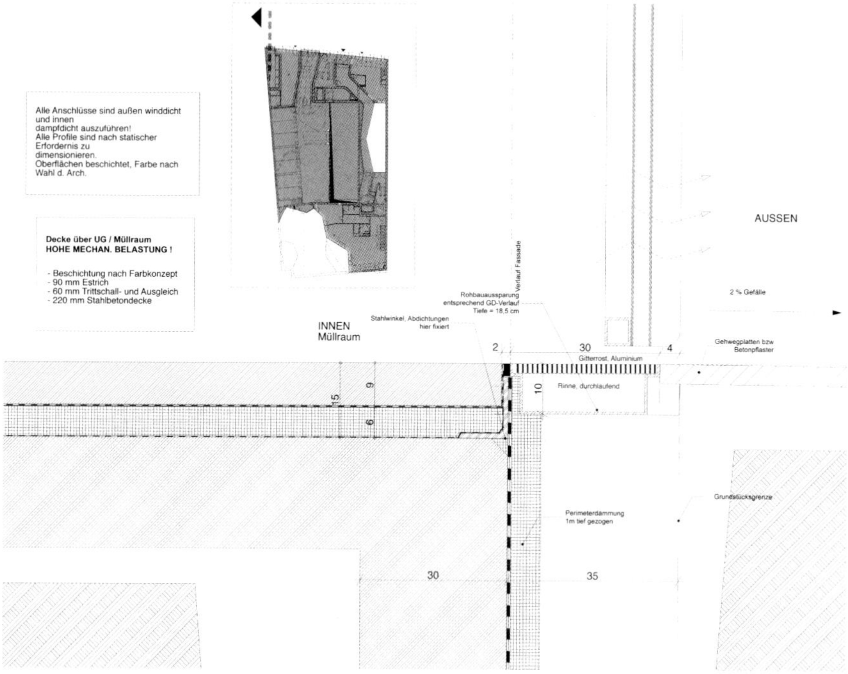

Detail 2 细节图 2

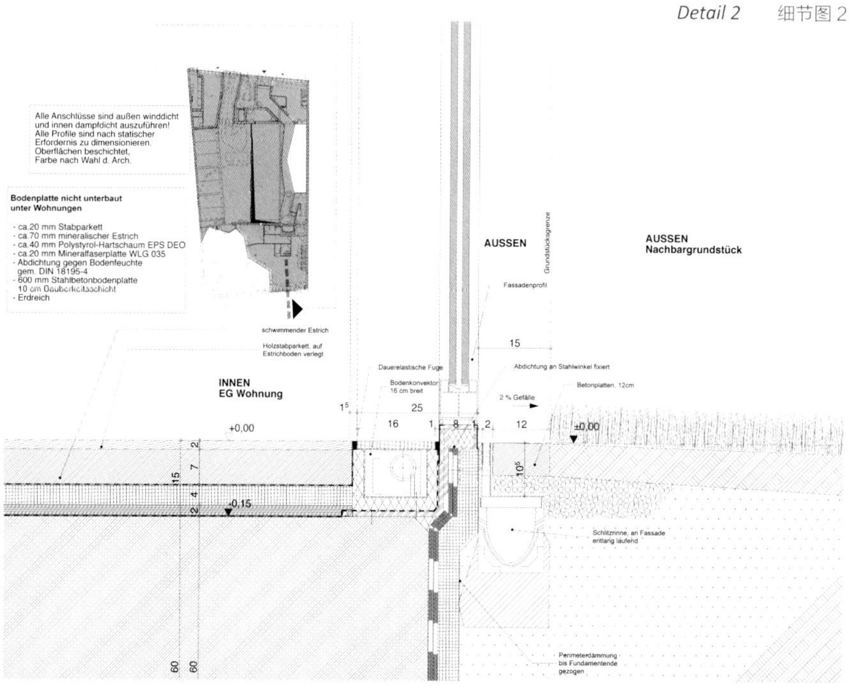

Detail 3 细节图 3

Intefeel, Osaka, Japan

Location: Osaka Japan
Architect: Tsubasa Iwahashi Architects
Site Area: 1,269.35 m^2
Gross Floor Area: 4,281.12 m^2
Number of Floors: 5
Photographer: Takumi Ota

日本大阪 Intefeel 公寓

地点：日本大阪
设计公司：岩桥翼建筑事务所
占地面积：1 269.35 m^2
建筑面积：4 281.12 m^2
层数：5
摄影师：太田拓实

Surmount 突破点

VIEW 视野

FACADE 立面

SPACE 空间

NATURE 自然

Nature is always next to the living. The richness of nature stimulates person's sense, and brings a lot of pleasure. We will be able to obtain energies and a relief manning original by being in the average of nature.

There are four seasons in Japan: in spring when flowers are in full bloom, fresh green energies overflow; in summer there is strong sunshine and shower falls; autumn is a harvest season; in winter when snow wraps towns, people have expectations that new season would rise. Various expressions are shown to us in all seasons.

This site is adjacent to the Osaka bay. There is a mountain range for the east, and it is located in a natural, rich environment that can overlook the sea for the west. The building is composed as a complex facility for the senior citizen. The cafe and the clinic opened also by people in the region are set up at a grand level.

The space volume invaded among those projects the expression onto the outside wall. The façade on the mountain side gives the mountain range that soars on the east side and steepness in the homage doing and the surface of a mountain has been given to the expression. The façade on the sea side tells the expression of the repeated wave to the outside wall.

An open mouth cut into ten walls is a design that takes the spectacle and the natural light adjoined while securing privacy with the tenant of standing next internally as much as possible. Nature that the living is soon in by the side is extracted, and the beginning of the extent is felt pure directly. Then, the people who live will find his favorite whereabouts between the building and nature. It is enriched while drawing close naturally on immediate side and it is spent.

IN FEEL

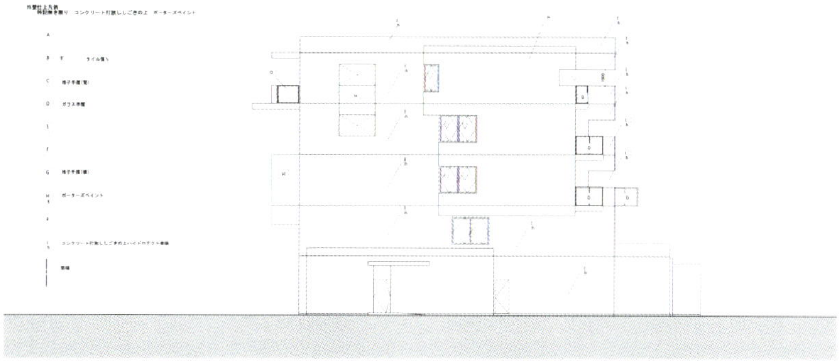

Northeast Elevation Plan 东北立面图

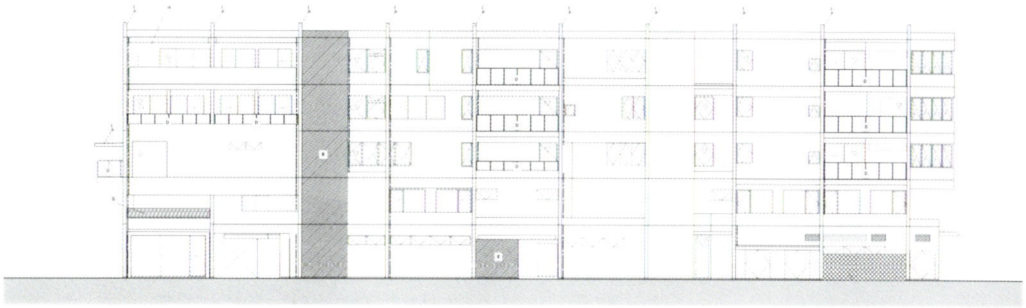

Southeast Elevation Plan 东南立面图

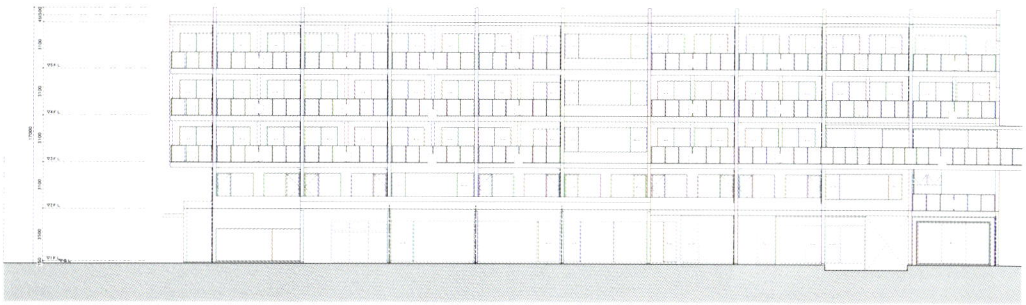

Northwest Elevation Plan 西北立面图

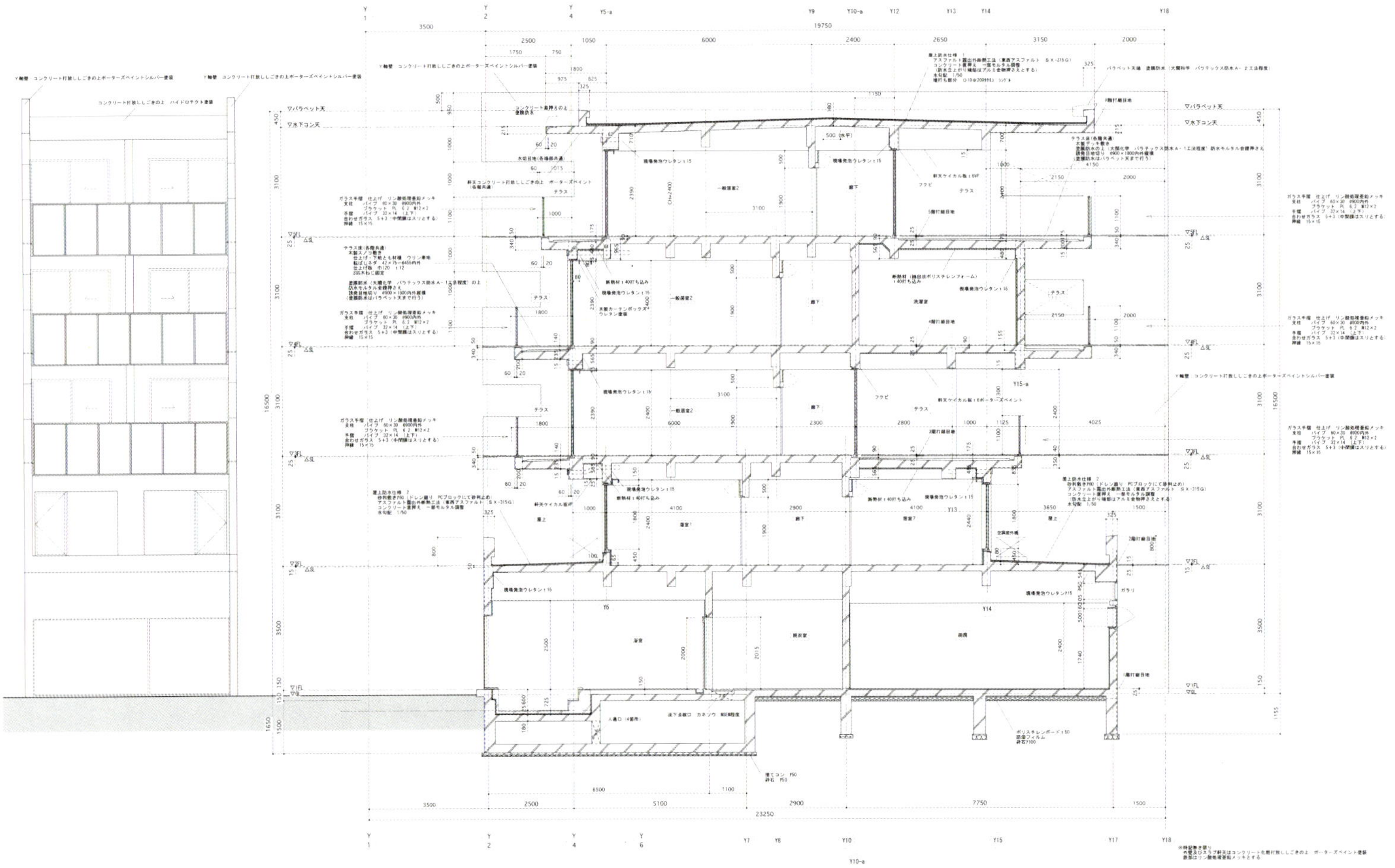

Section Plan 剖面图

自然总是在我们左右，丰富的大自然刺激着我们的感官，同时带来许多乐趣。在自然中我们总能获取能量并释放心情。

日本四季分明：春天百花齐放，绿意盎然；夏天阳光强烈，阵雨时至；秋天则是丰收的季节；而冬天到来时，市镇将被冰雪包裹，并在寒冷中期待着新季节的来临。不同的季节展现着不同的风情。

这座建筑是一栋供老年人使用的综合性公寓楼，位于大阪附近，东边是山脉，西边可以俯瞰大海，处于一个自然、富饶的环境中。公寓一楼设有本地人开的咖啡馆和诊所。

体量的回缩也体现在建筑外墙上面。山边的立面体现了东边山脉的高耸外形和陡度。海边的立面则在外墙上体现着汹涌不断的波浪。

这十面墙采用了开口设计，在尽可能地保护了租户隐私的同时，让住户享受到更好的景色和光照。这里的生活提取了自然的精粹，让人直接感受到大自然的纯净。居民们将会在建筑与自然之间找到归宿，在接近自然的同时充实自己。

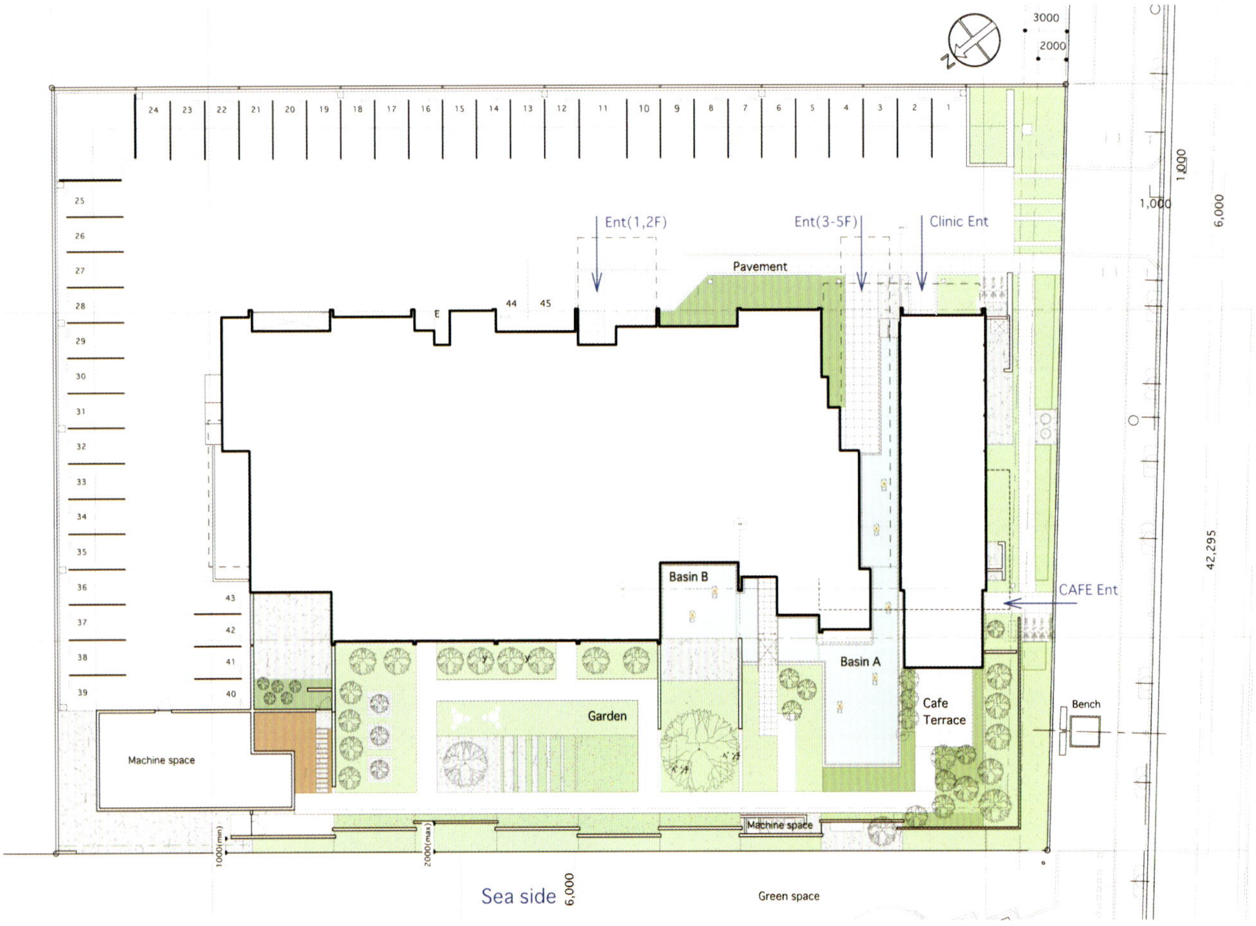

Green Space Plan 绿化区域图

FEATURE 特点分析

Space

Construction produces the space with a concrete wall construction. The role to make the scene of the living a division also accompanies 10 walls that run parallel postponed from east to west. And, the scene of a living consecutive as the mass of the joined space is invaded between the walls that run parallel, and the ties are built. The space of the living with individuality is set up in irregularity and a complicated volume. The consideration that can be faced various nature seen in the place is performed to the space.

空间

建筑用混凝土墙来划分区域空间。10 面平行墙自东向西排列，划分了连续的生活场景。由多个被平行墙隔开的空间组成了连续的生活空间，把住所紧紧联系在一起。个性化的生活空间则建立在一个不规则的复杂体量中，也让空间能面对更多样的自然景色。

5th Floor Plan　5 层平面图

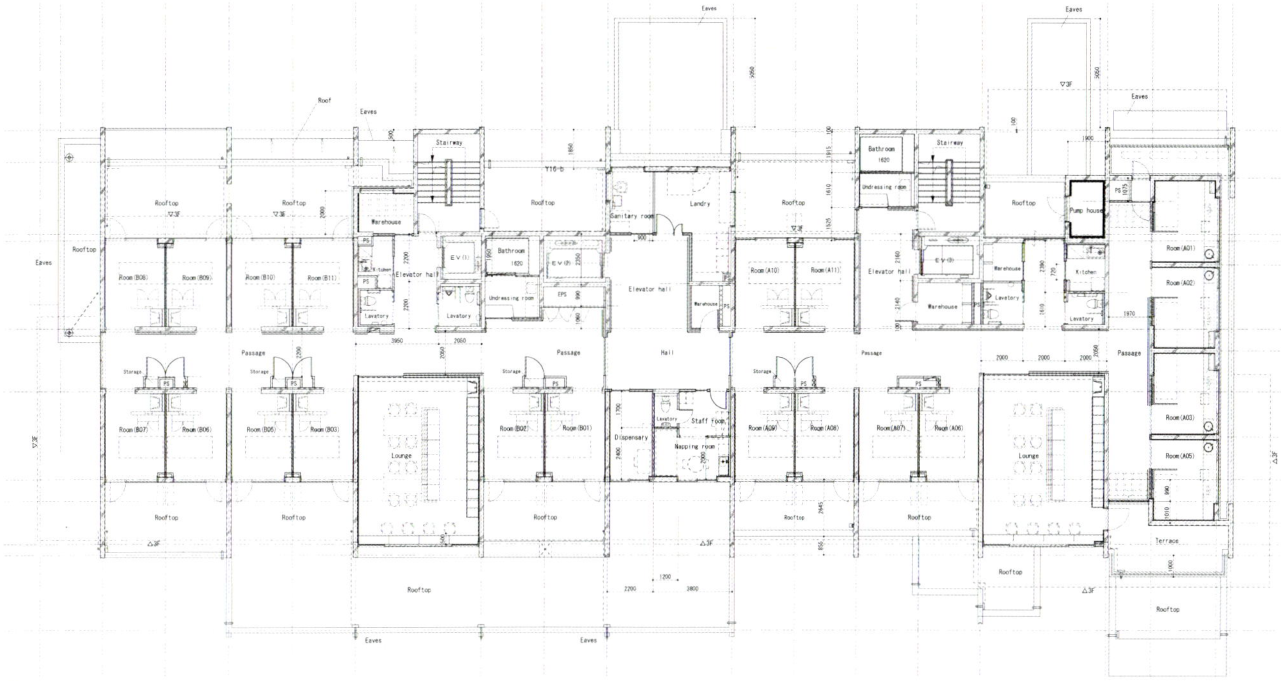

2nd Floor Plan　2 层平面图

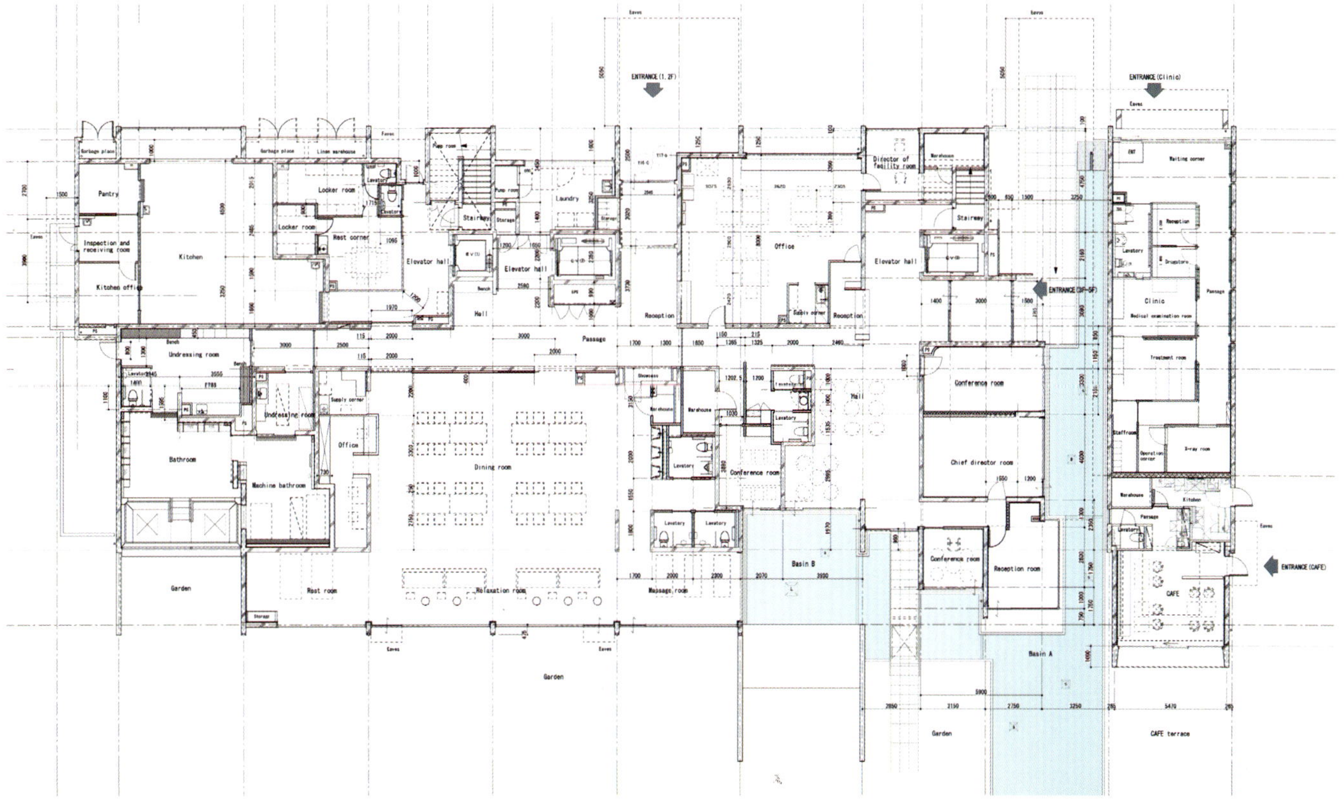

1st Floor Plan　1 层平面图

Boulogne D2, Paris, France

Location: Paris, France
Architect: 2/3/4/ Architecture
Site Area: 3,850 m²
Gross Floor Area: 31,148 m²
Height: 34.90 m
Number of Floors: 11 (the main building); 9 (the 2nd building)
Plot Ratio: 4
Photographer: Fernando Guerra

法国巴黎 Boulogne D2 公寓楼

地点：法国巴黎
设计公司：2/3/4/ 建筑事务所
占地面积：3 850 m²
建筑面积：31 148 m²
高度：34.90 m
层数：11（主楼）；9（裙楼）
容积率：4
摄影师：费尔南多 · 格拉

Surmount 突破点

BALCONY 阳台

VIEW 视野

SHAPE 造型

DAYLIGHTING 采光

The project is composed of 2 spatial squares opposing each other across a large shared central garden:

The first square begins at the southwest corner along the Seine and then returns along the large traversal border where it eventually reaches the park. This square is divided into two parts: a stone building 34.9 m high facing the Seine and a linear building along the traversal road.

The opposing square, with its building of shorter stature, containing the social rental housing units, runs alongside the park to the northeast, returning along a small bordering traversal road shared with an office building.

The volumes conceived meet not only the need of creating a visual relationship between the park and the Seine, but also of creating a similar visual relationship with the neighboring office buildings. The alternation of high (GF+10) and low (GF+8) volumes, and the presence of continuous terrace-deck spaces facing the Seine impart a rhythm of continuity to the front of the Quai de Stalingrad.

A gallery completely free from the structure lies fitted with an underside of 5.5 m along the entire length of the quay, with an accompanying depth of 2.8 m inwards from this alignment with the quay.

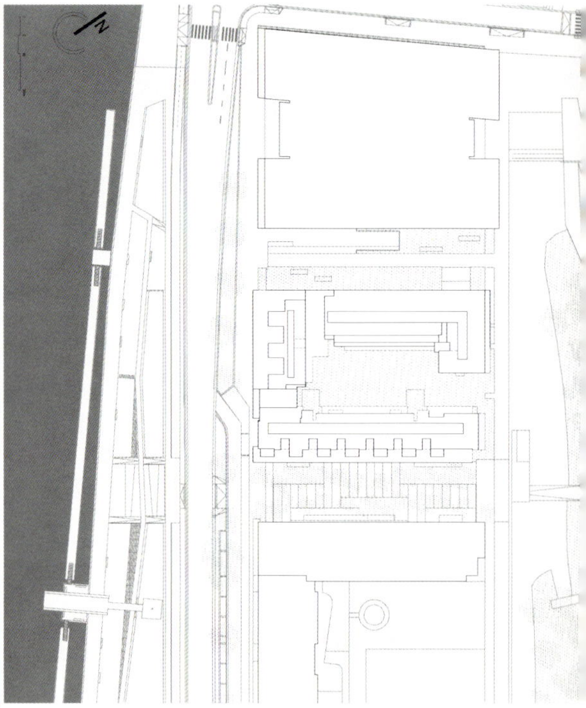

Site Plan 总平面图

The extremities of the two opposing squares converge facing the park. The urban islet-block benefits specifically here from a large opening of around 10 m in height that allows the establishment of a correspondence with the vegetation present at the heart of the plot.

On the side of the large traversal road, the linear building faces the park along all of its height, where a cantilevered volume of double height crowns it.

On the side of the smaller traversal road, the square, composed here of its building of social rental housing, is raised up two floors, which in consequence allows the park to "slip in through" and enlarge this fracture between the two buildings.

The base (composed of two floors) is broken up architecturally by the individual entryways of the housing units.

The central structural body (of 6 floors) is enlivened by large, deep balconies arranged in a dynamic pattern of staggered placement, which in turn creates side lateral sections of extended height. The remaining sections of simple height are protected by a glass wall, which provides shelter from the wind, while granting unobstructed views.

The attic, punctuated by a series of 7 duplex volumes, crowns the building.

The building running alongside the small traversal division leans up against the stone edifice. Its low height allows plenty of natural light to pass through to the private road. The base is at 80 cm remove in relation to the structural body of the building. This body is above all characterized by the regular lines of its continuous terraces, along which large shutters in exterior alignment have been placed randomly.

The private garden constitutes the heart of the project and allows the appreciation of the whole of the architectural structure from a single and singular perspective: the transparency and sobriety of the constructed volumes, in counterpart to the vegetation it includes.

All the large apartments of the project are built with multi-exposure and orientation in mind, with the exception of a series of 4 room units at the south of the building. These two residences are developed as duplexes so as to interlock together laterally. Their reduced thickness and length, as well as the absence of corridors that result from their interlocking, allow a full amount of natural light to pass through even to the depths of the space. The generous amounts of double height built into the entryways and at the balconies grant these duplexes a particularly domestic and private scale, somewhat after the image of a small village home.

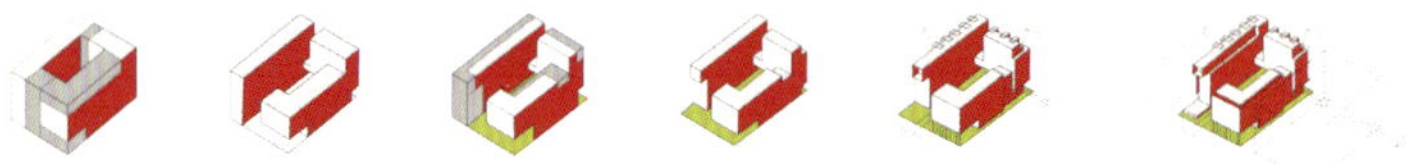

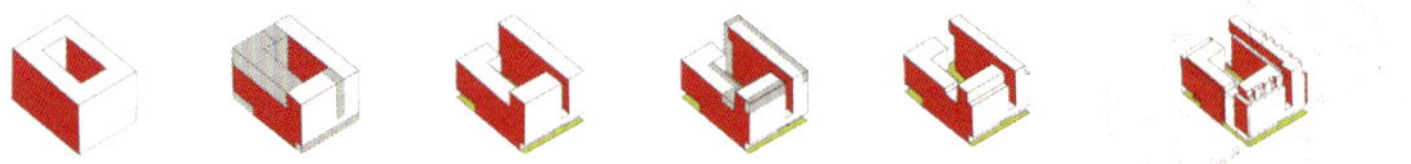

Volume Development Scheme 体量演变图

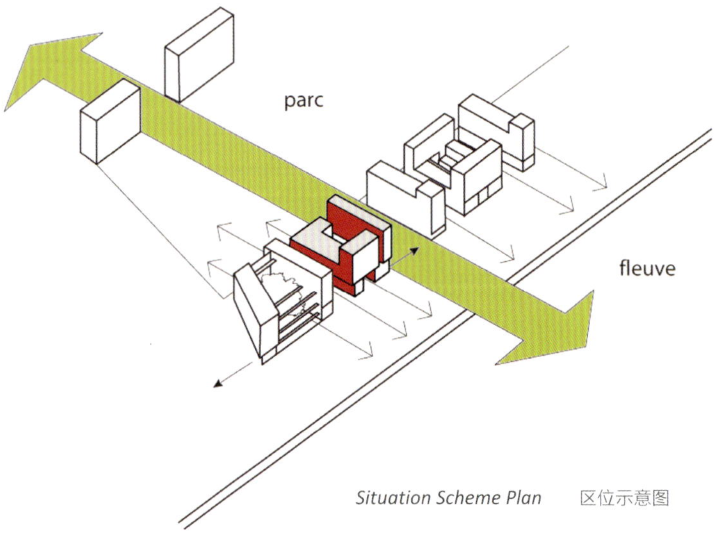

Situation Scheme Plan 区位示意图

RESTAURANT

项目包括两个相对而立的空间广场，中间横穿一个大型公共中央花园。

其中一个广场位于塞纳河的西南角，沿着河边延伸，横穿边界，最后连接到公园。这个广场分为两部分，包括一栋面对着塞纳河的 34.9 m 高的石头建筑，以及一栋穿越公路沿线的线性建筑。

对面的另一个广场较为狭窄，建有社会福利出租房，沿着公园延伸到其东北面，穿过一条小边界公路返回。公路上还有一栋办公楼，与广场一起装点着公路。

建筑的体量设计不仅要满足公园和塞纳河之间的视觉联系，还要和四周的办公大楼产生类似的视觉联系。两栋分别为 10 层和 8 层高的建筑和面向塞纳河的连续露台，赋予了前方码头一种连续的韵律。

码头底部还安排了一个独立画廊，内部根据码头的定线高 2.8 m，沿着 5.5 m 长的码头一直延伸。

这两个对立的广场的末端交汇并面对着公园。从周围 10 m 高的地方都能看到这片开阔的城市小岛，中央的绿色植被也更显和谐。

大边界路的一侧是一栋线性建筑，其中一个面对公园的立面顶部采用了双层悬臂结构。

小边界路一侧是社会福利出租房，位于两层高的广场上，使公园看起来如同穿行其中一般，也使两栋建筑间的空隙更宽敞。

两层高的地下室在建筑上被各个住宅单元独立的门廊所划分。

又大又深的阳台以一种充满动感的形式交错排列，使 6 层高的中央结构体更显生机活力，同时也使横向部分显得更为高大。其他一般高度的部分由玻璃幕墙保护，既能避风，又能让人欣赏到无限美景。

建筑的阁楼由 7 个一系列的双层结构体构成，傲立于楼顶。

建筑沿着小边界路的岔路延伸，斜对着一栋石头大厦，其高度较低，使自然光线可穿过它照进私人小路。建筑的基部与建筑的结构体相距 80 cm。建筑最大的特点是其连续露台的规则线条，大型的遮阳板根据外部定线随机放置。

项目最核心部分是其私家花园，从这里可以以一个最独特的视觉效果欣赏到建筑结构的整体；建筑体量的通透性和严谨性与其内部植物相得益彰。

除了建筑南边的 4 室公寓套间，项目里所有的大型公寓都考虑了多方位的光线和朝向。其中，这两套住宅被设计为复式套房，以便其横向相连，这导致了其厚度和高度的减少，缩小了走廊的空间，使自然光线可直接照到房间深处。大部分的门廊和阳台都建成双层高度，使这些复式套房更具自主性和私人空间，看起来更像是一座乡村小别墅。

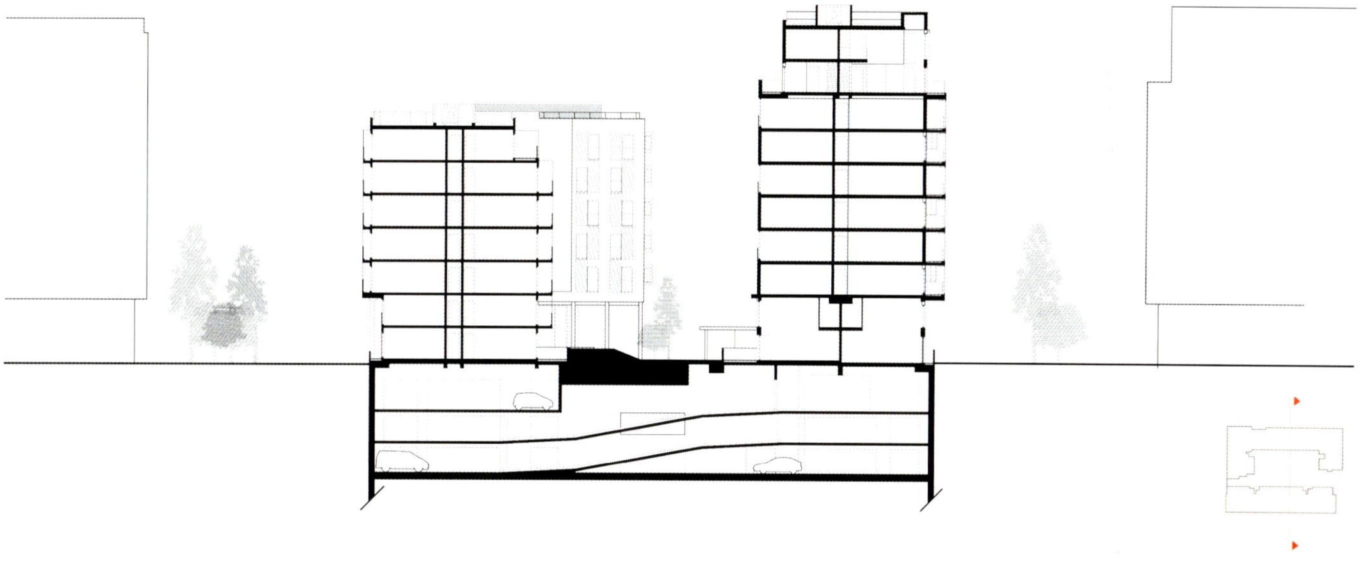

Section Plan 1　剖面图 1

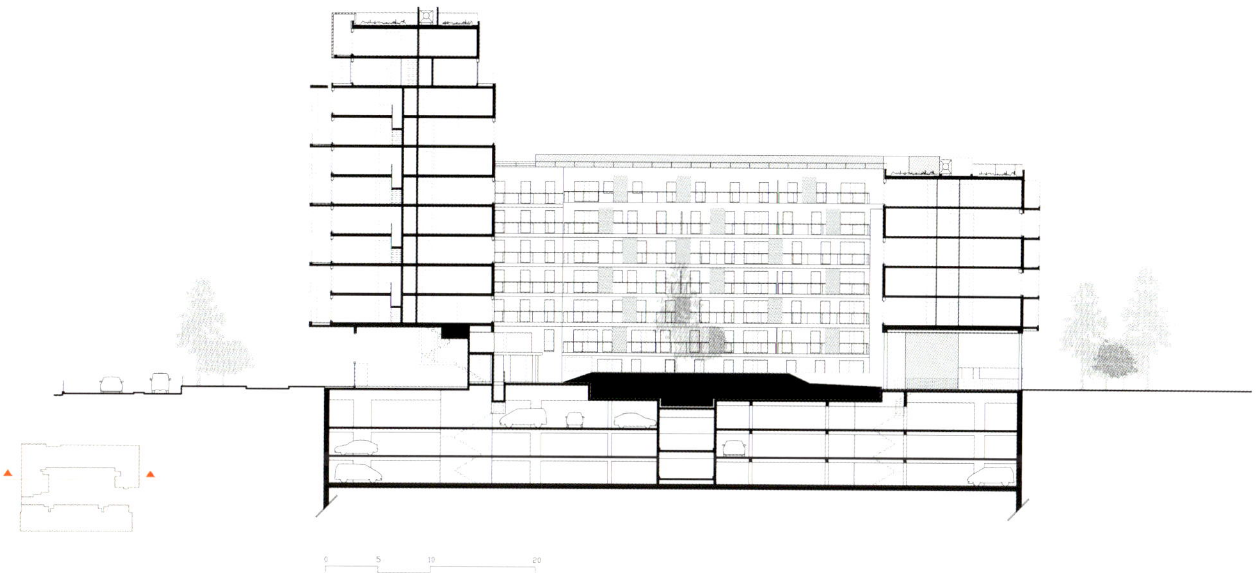

Section Plan 2　剖面图 2

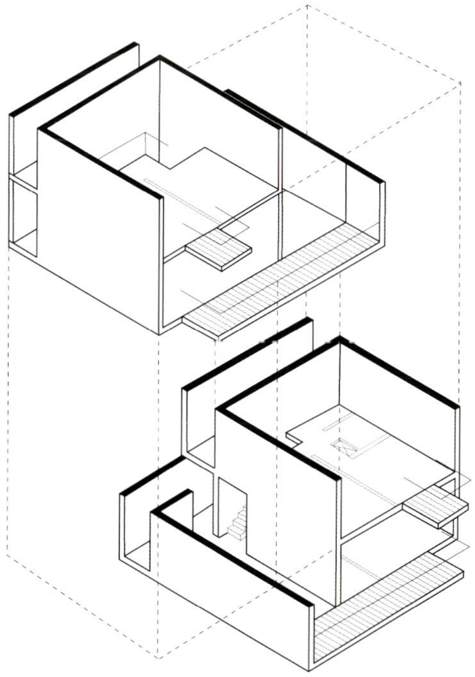

Structure Analysis Plan　结构分析图

Model　模型图

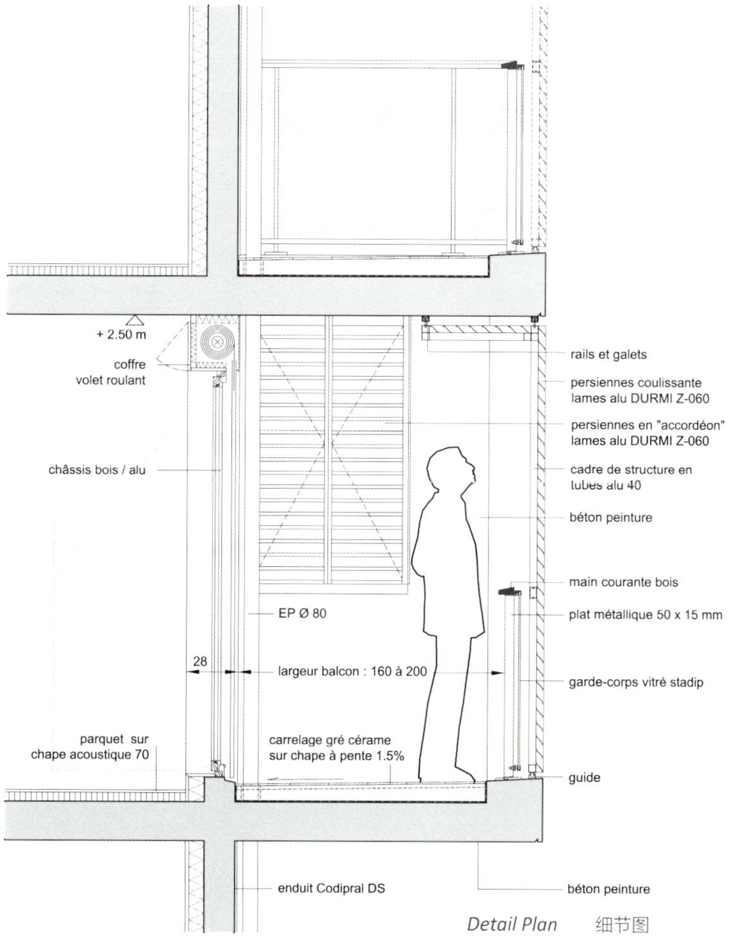

Detail Plan 细节图

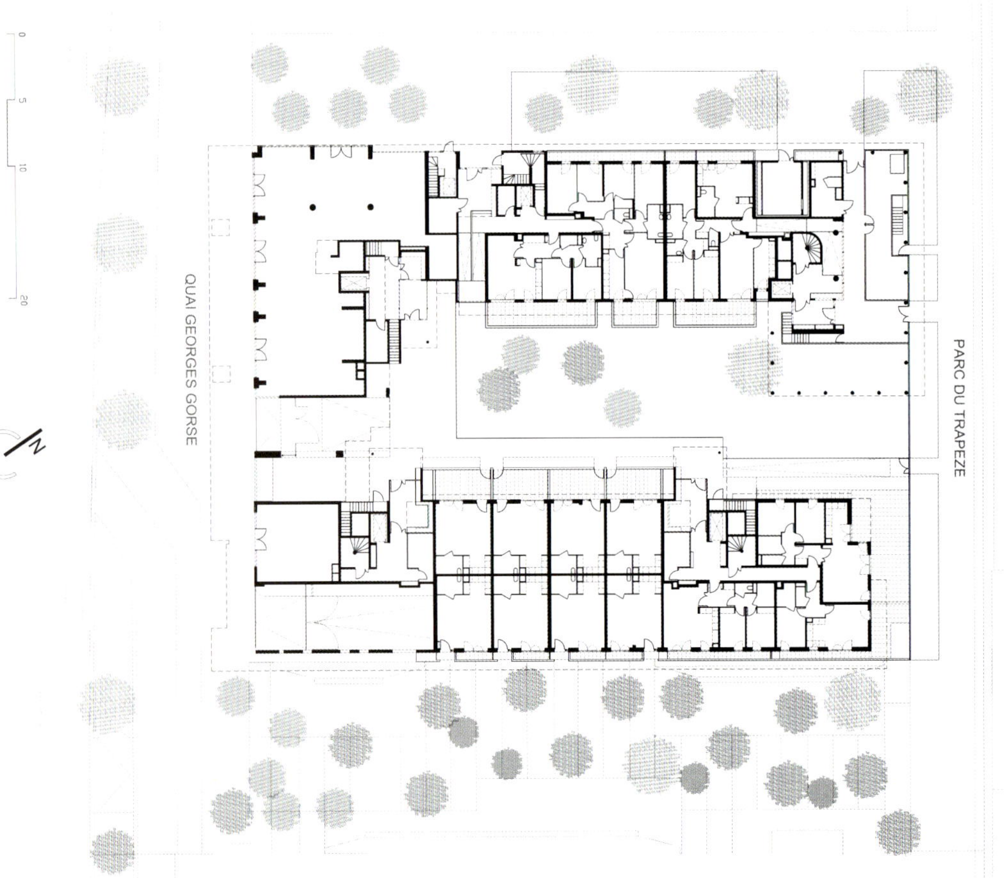

Floor Plan 楼层平面图

FEATURE 特点分析

View

One façade of the building faces the meandering river with open view and another one towards the community park. The ground floor is high and fully transparent. Almost all households is designed in two orientations, the rest single orientation households will face the park or the Seine River.

视野

建筑的一个侧面朝向蜿蜒曲折的河流，视野开阔，另一侧则朝向社区公园。建筑首层的层高较高，并完全透明，多数住宅设计有两个朝向，而为数不多的单一朝向的住宅则朝向公园或塞纳河。

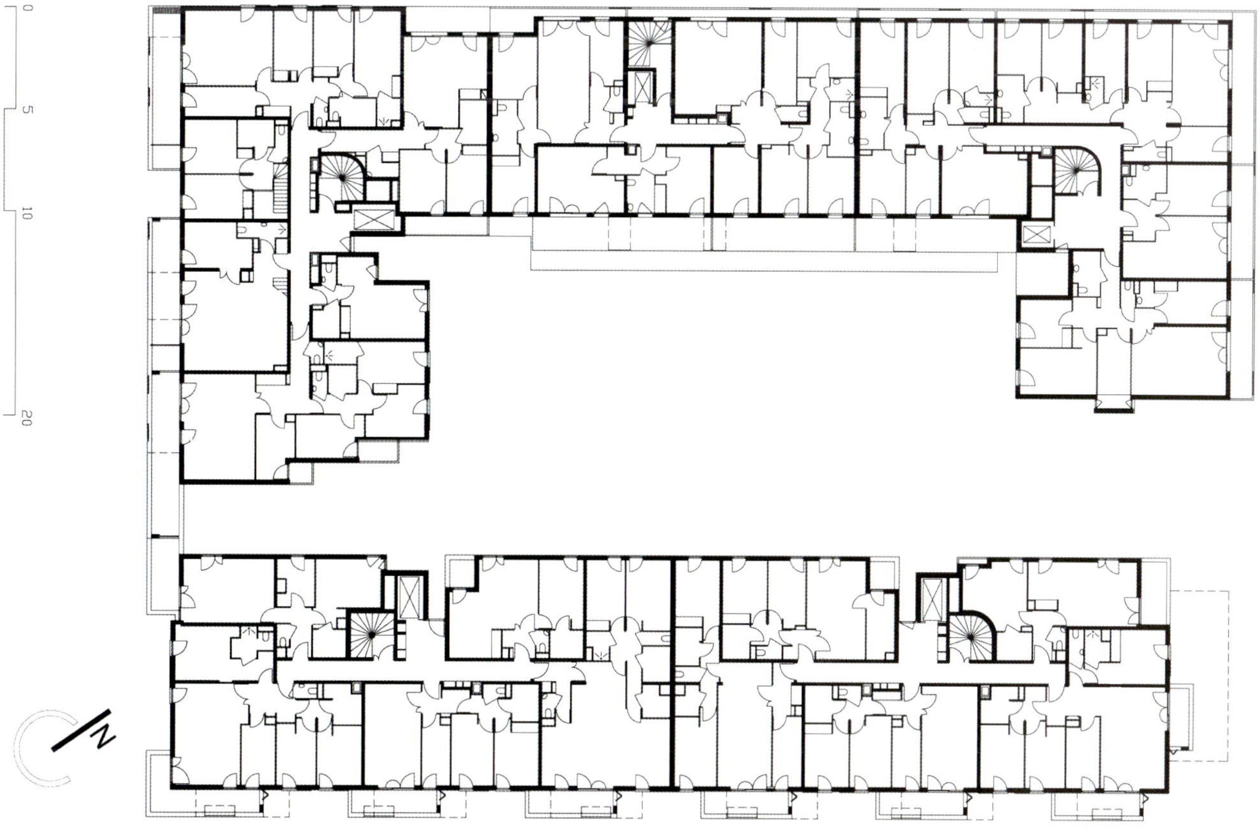

Floor Plan 楼层平面图

Residential Building Plokite Split, Croatia

Location: Split, Croatia
Architect: Arhipolis Achitects
Site Area: 3,500 m²
Number of Floors: 7
Photographer: Arhipolis Archive

克罗地亚斯普利特市 Plokite 公寓楼

地点：克罗地亚斯普利特市
设计公司：阿里波历斯建筑事务所
占地面积：3 500 m²
层数：7
摄影师：Arhipolis Archive

Surmount 突破点

SUSTAINABILITY 可持续性

SHAPE 造型

MATERIAL 材料

STRUCTURE 结构

This building is located into the existing context of medium-density buildings in a quiet part of town.

Facility has been designed on the location of the existing building which is demolished. The intention was to fit the new object with its volume and appearance harmoniously into the existing context and to create a better environment for existing buildings and neighboring school.

This 4 storey building includes 20 housing units and the underground parking. Ground floor apartments have private gardens and balconies which are elevated in relation to the road that allows them privacy. All other flats also have large balconies that stretch the full length of the building.

Most of the used materials can be recycled and the facility uses solar energy for water heating. Base of the house has perforated metal façade while the rest of it is plastered and painted.

这栋建筑位于小镇上一个安静的角落里，隐藏在众多中等密度的楼宇中。

建筑设施的设计则参照了当地被拆除的建筑，使新建筑的体量和外观与周围现有的环境更和谐，并为周围的建筑和附近的学校创造一个更好的环境。

建筑高 4 层，共有 20 个公寓单元和 1 个地下停车场。首层的公寓房配有私人花园和阳台，比外面的道路高出许多，以保护住户的隐私。其他公寓房也设有大阳台，沿着建筑的同等长度延伸着。

建筑使用的大部分材料都是可以循环利用的，并配备了太阳能热水器以提供热水。建筑的基地采用穿孔的金属板幕墙，其他部分则采用了抹灰和油漆立面。

FEATURE 特点分析

Shape

Conceptually it consists of horizontal elements – "boxes" that are mutually slipped aside and form a dynamic composition of volumes. This concept has been consistently applied in all aspects of the building such as details and polychromy.

造型

从概念上讲，建筑是由一系列横向元素组成的——一个个"箱子"错落分布并向外突出，打造出这个体量的视觉动感。这个概念一直延伸到建筑各部分的细节和色彩装饰。

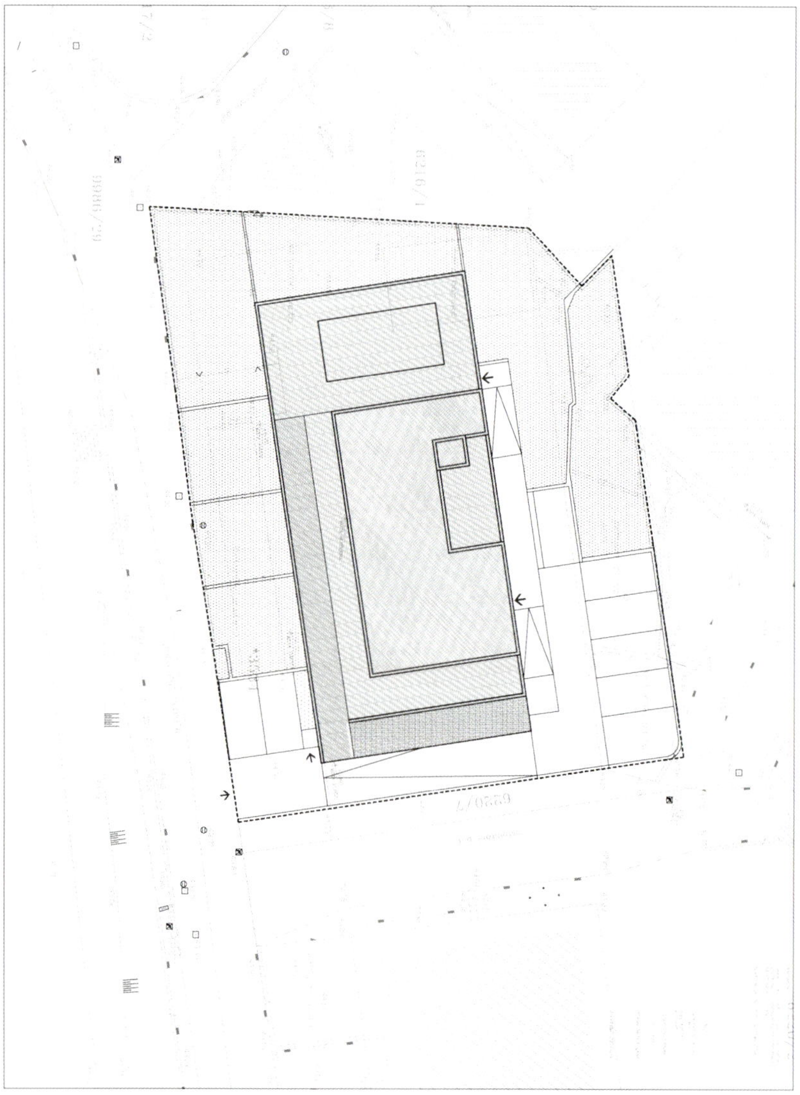

Site Plan 总平面图

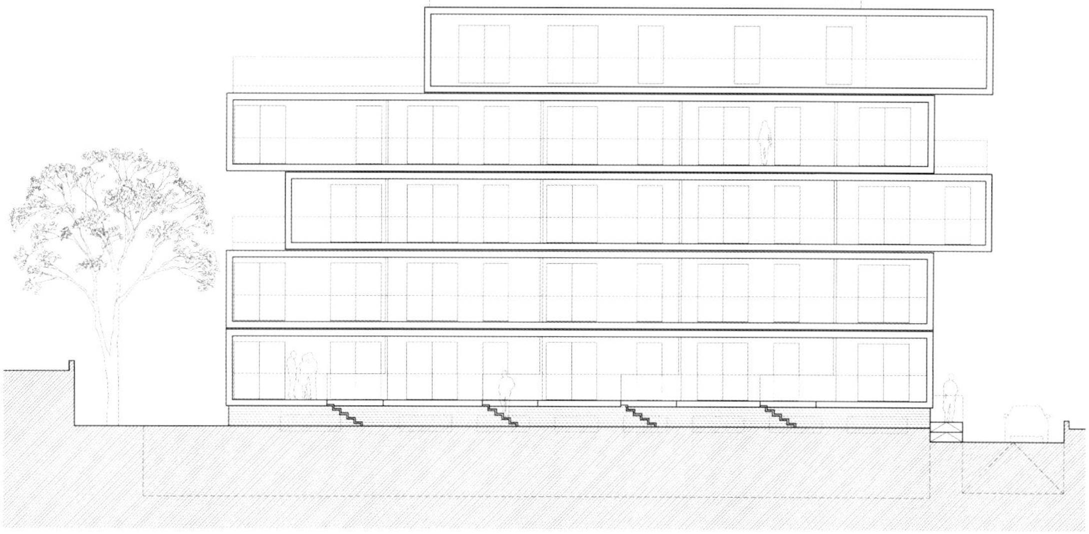

South Elevation Plan 南立面图

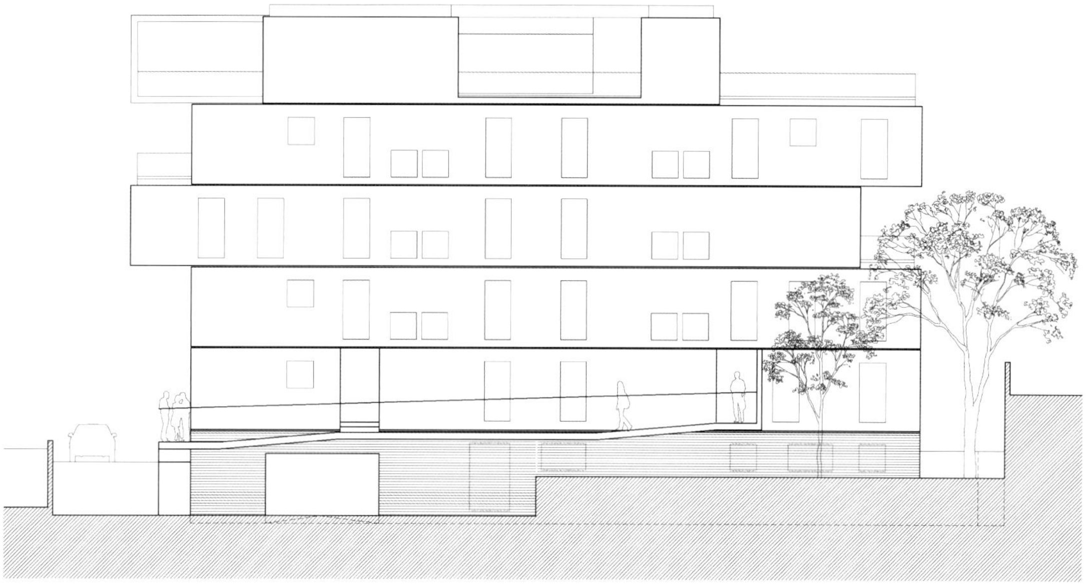

North Elevation Plan 北立面图

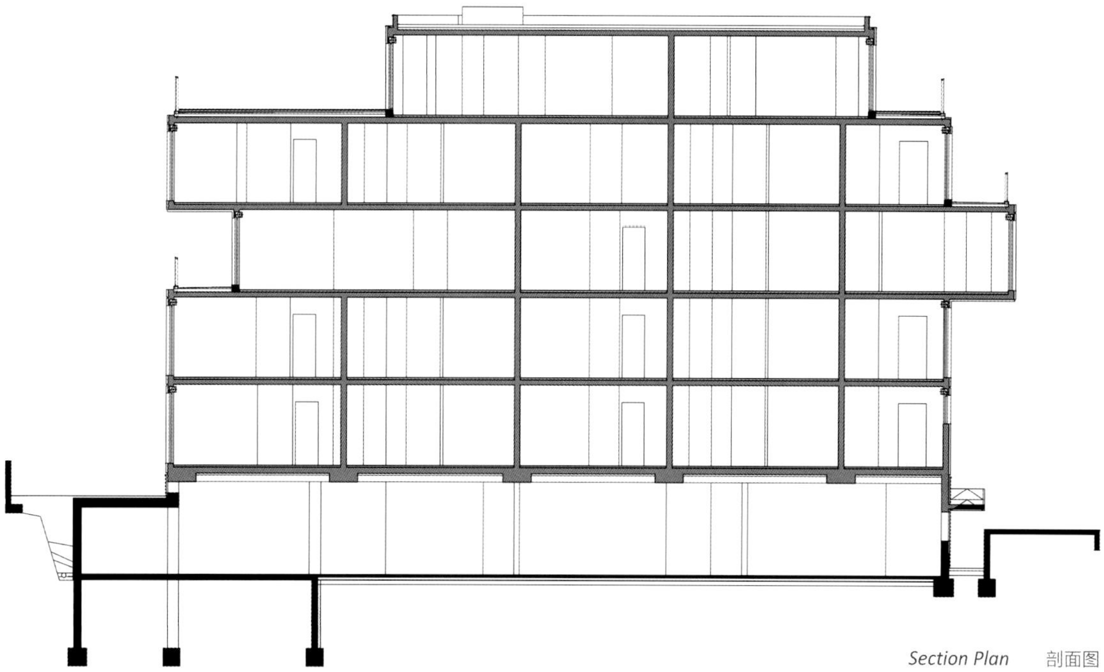

Section Plan 剖面图

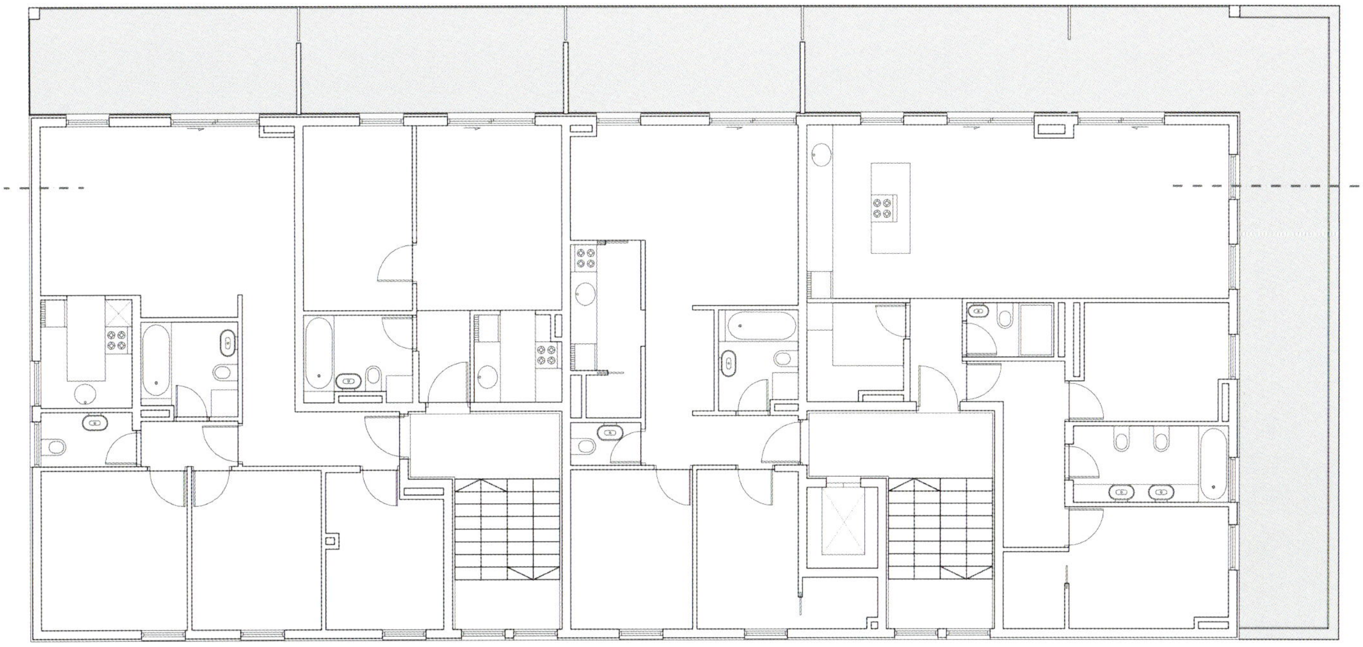

Floor Plan 楼层平面图

Lyon Dynacité, Lyon, France

Location: Lyon, France
Architect: 2/3/4/ Architecture
Site Area: 2,432 m^2
Gross Floor Area: 6,091 m^2 (the lower building), 2,520 m^2 (the higher building)
Height: 18.5 m (the lower building), 26.65 m (the higher building)
Number of Floors: 6 (the lower building), 9 (the higher building)
Plot Ratio: 3.54

法国里昂市 Lyon Dynacité 社会福利公寓房

地点：法国里昂市
设计公司：2/3/4/ 建筑事务所
占地面积：2 432 m^2
建筑面积：6 091 m^2（低层建筑）；2 520 m^2（高层建筑）
高度：18.5 m（低层建筑）；26.65 m（高层建筑）
层数：6（低层建筑）；9（高层建筑）
容积率：3.54

Surmount 突破点

FUNCTIONALITY 功能性
FACADE 立面
MATERIAL 材料
PRE-CAST 预制

The project puts up two contents: a housing estate and an emergency accommodation structure, and is designed as a consistent whole to promote diversity.

The metal floors blend in the sky by changing, reflecting and creating a shredded skyline. Thereby, the repetition of the public housing design ruled by functionality and standards building techniques is not visible from the outside.

Flats design is guided by three main issues:

1)Entrance halls are set apart to maintain intimacy; concealed sliding doors give access to the living space.
2)Openings are given a maximum size considering the flats layout, spandrels have been recessed to let light in.
3)Prefab elements will accomodate bathrooms, custom-designed but prefabricated to ensure perfect quality of execution.

The qualities of urban spaces basically reside in the way building and public space are connected. The design of the ground floor, mineral and dark concrete, refers to the street appearance. It opens to the street through a sequence of porches, halls and overhangs. Thereby, the main volume of the building is raised for the most part of its extension, and opens onto the block courtyard at east, allowing the perception of strong permeability between the green road and the inner garden. The dark color of the bottom part of the façade emphasizes by contrast the views to the exterior from the center of the block. The project gives to its inhabitants a place to live at the same time intimate, strongly environmental and functional. It opens onto the city and the river through an easy path.

本项目共有两部分，包括住宅楼和紧急避难住宅楼，两者设计成一个协调的整体以提高多样性。

金属地板反射着光芒，不断变幻，并和天空融为一体，创造出一段细碎的天际线。功能化和标准化的建筑技术使该建筑从视觉上完全摆脱了一般公共住宅重复的设计规则。

公寓的设计遵循了三条原则：第一，入口大堂独立设置以保护隐私，使用隐蔽式滑动门作为通往生活区的出入口；第二，公寓的布局让空间得到最大化利用，拱肩往里凹进让光线可以照射进来；第三，公寓设有许多专为客人设计的、包括浴室在内的预制部分，以保证公寓的高品质。

建筑和公共空间的联系往往决定了城市空间的质量。根据街道的外观，该项目首层的设计采用了矿石和深色的混凝土，依次通过门廊、大堂和屋檐面向街道展现开来。因此，建筑的主体量大部分往外伸展，与东边的庭院相通，使绿色的道路和内部花园呈现出一种强烈的相互渗透感。立面底部深色的部分通过对比强调了从建筑中心到外部的景观。项目给住户提供了一个亲切、环保和多功能的生活空间，同时开辟了一条城市与河流之间的便捷通路。

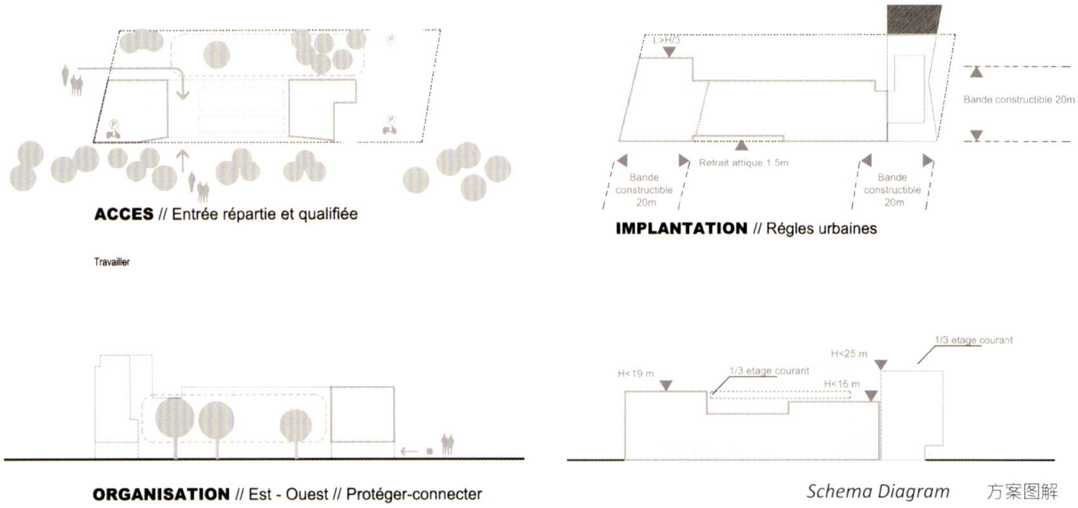

Schema Diagram 方案图解

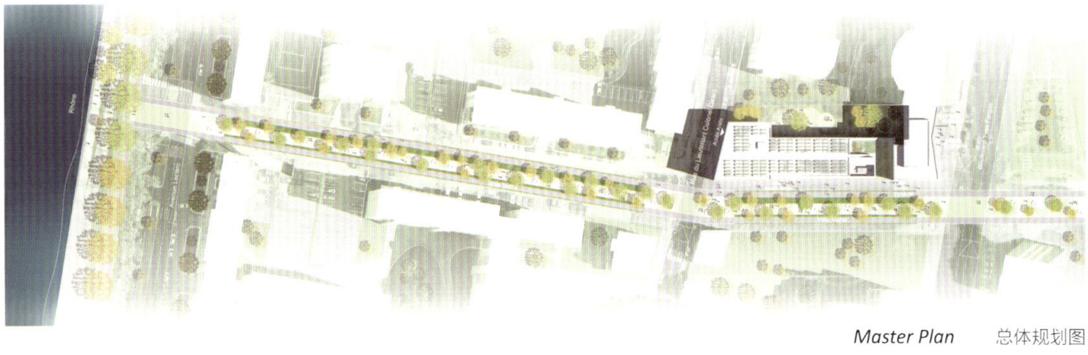

Master Plan 总体规划图

Elevation Plan 立面图

1st Floor Plan 1 层平面图

Section Plan 剖面图

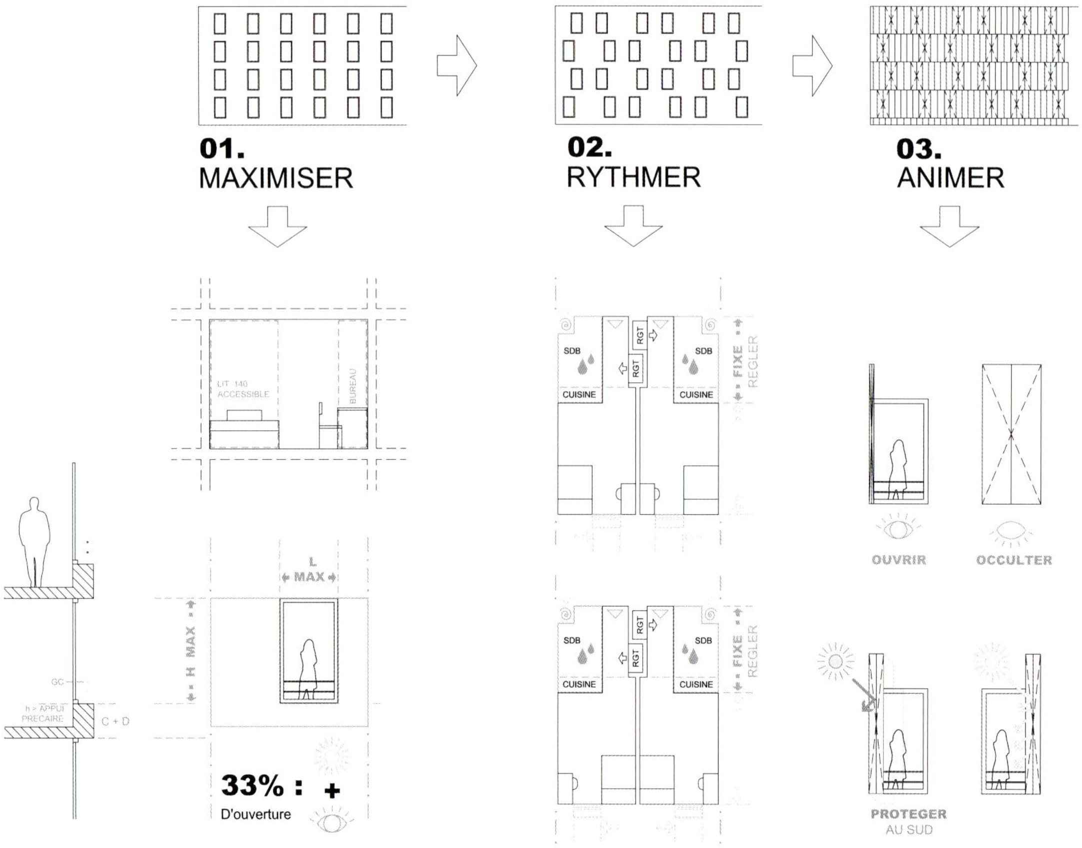

Elevation Analysis 立面分析

FEATURE 特点分析

Façade

The shifted windows of level to another animate the façade by creating a random pattern. Satin-finished and anodized aluminum shutters reflect the occupancy of the building, their aspect and color will vary with light.

立面

水平滑动的窗户通过随机设置的模式，赋予了建筑立面生机与活力。抛光处理的阳极电镀铝制百叶窗根据居住情况设置，随着光线变幻出不同的外观和色彩。

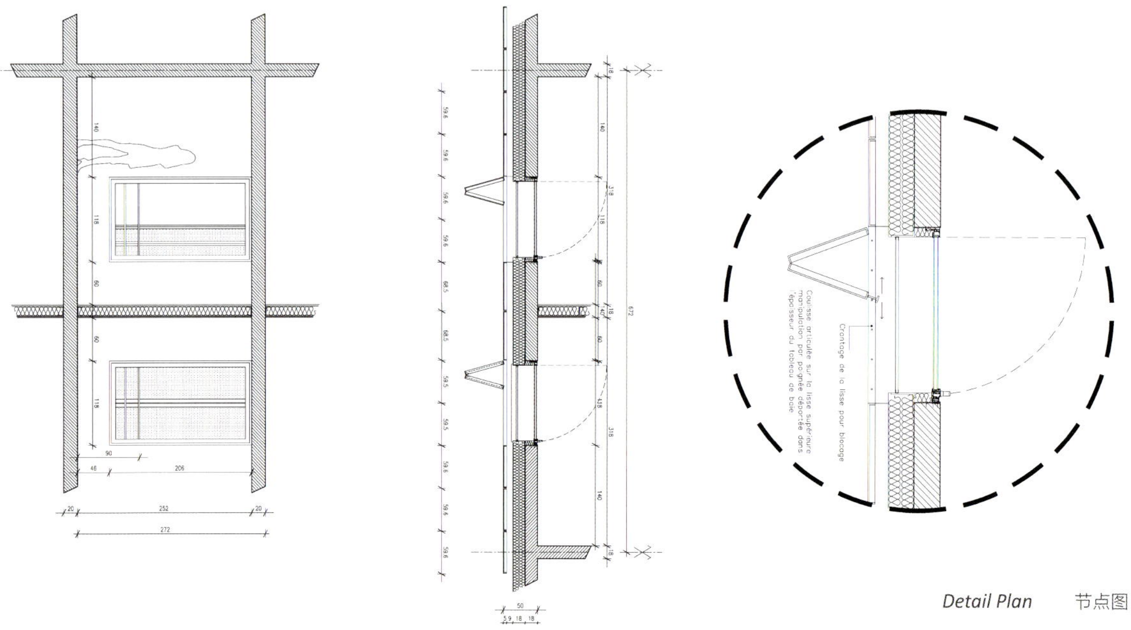

Detail Plan 节点图

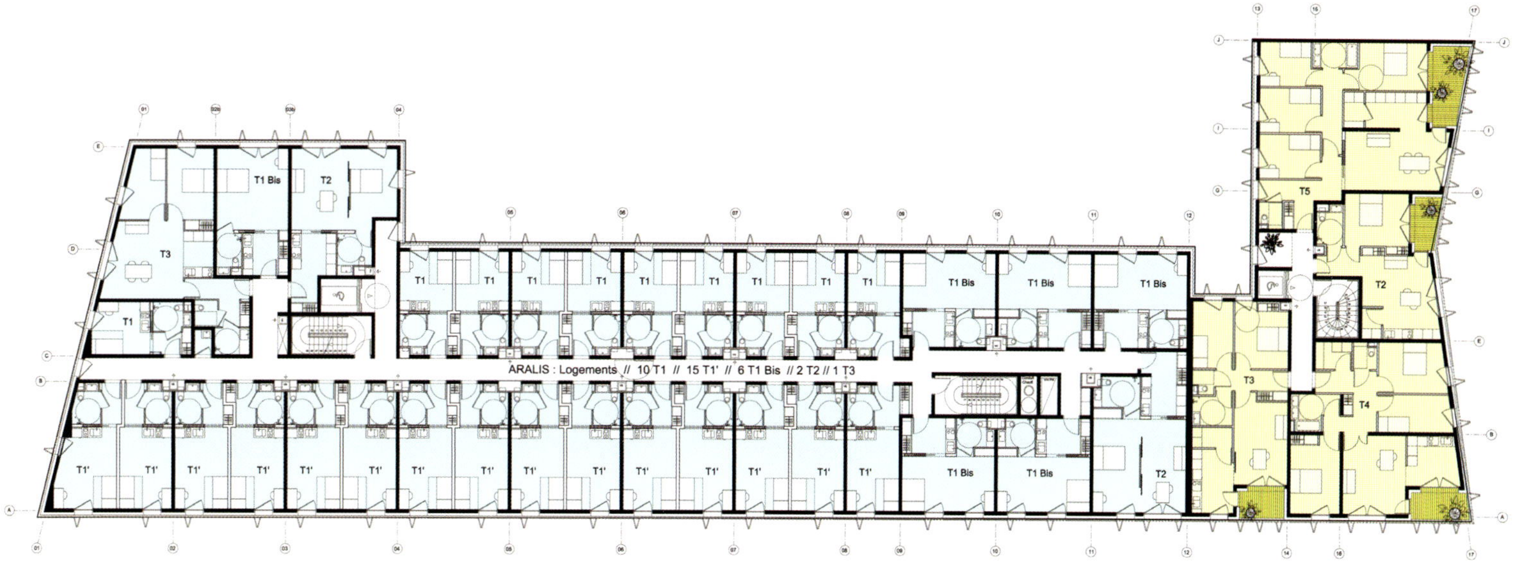

ARALIS : NIV. 01 à 03
NIV. 04 // + xxx.xx

DYNACITE : NIV. 01 & 02
NIV. 01 // + xxx.xx

Floor Plan 楼层平面图

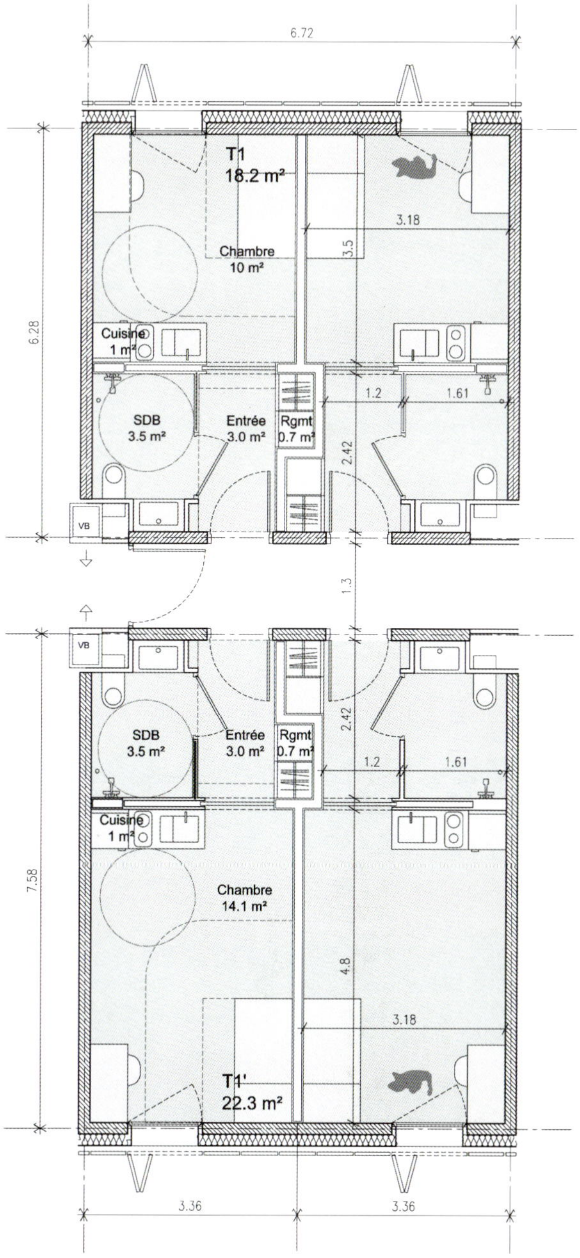

Details Plan 节点图

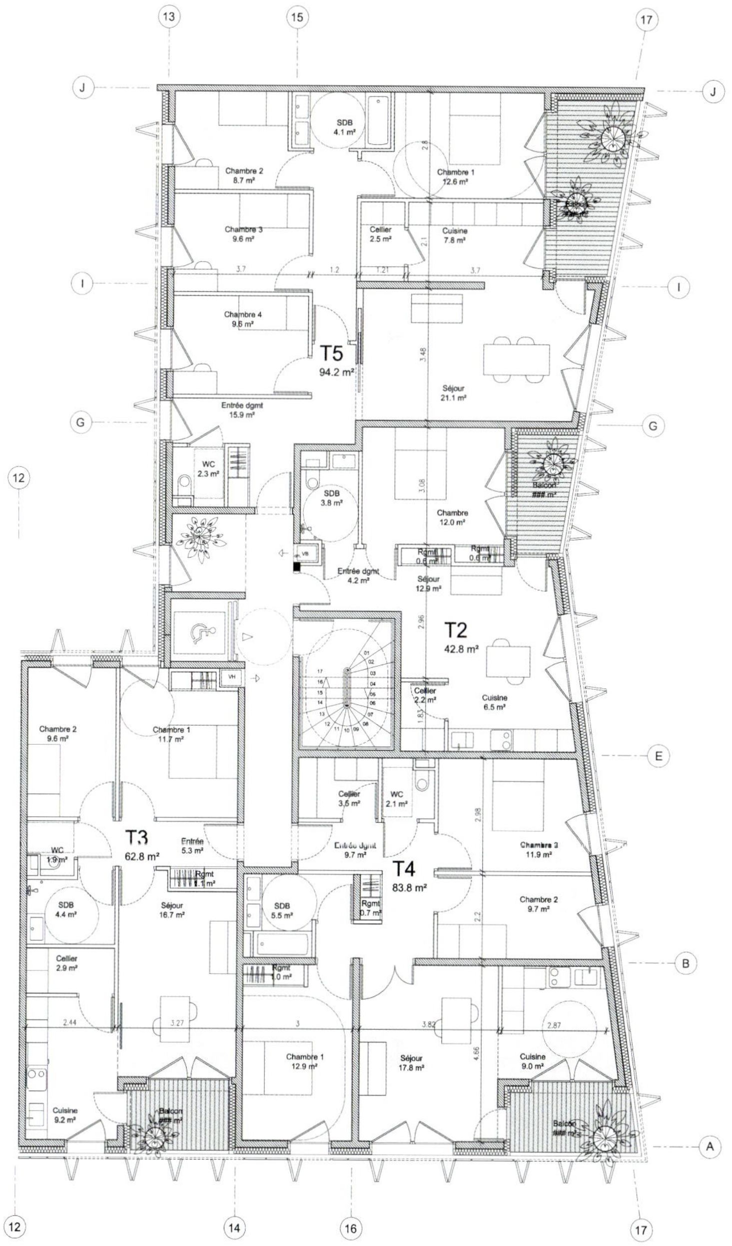

Floor Plan 楼层平面图

Social Housing in Angers, France

Location: Angers, France
Architect: Studio Bellecour
Site Area: 8,042 m^2
Number of Floors: 5
Photographer: Nicolas Borel, Philippe Caumes & Willy Berré

法国昂热市社会住房

地点：法国昂热
设计公司：贝勒库尔工作室
占地面积：8 042 m^2
层数：5
摄影师：尼古拉斯·雷波尔、菲利普·乔姆斯、威利·贝尔

Surmount 突破点

BALCONY 阳台

SHAPE 造型

ECONOMIC 经济

CITY LANDSCAPE 城市景观

The project sets within a mission of movement and punctuation. With this framework, a typological approach seeks to combine the urban scale of the street with its surrounding area. The project maintains the necessary "boundaries" between the building and the street while taking into account the area's density. The result produces an intimate residential atmosphere while maintaining the autonomy of the individual units, which break away from the standard "concrete block" and its the negative stereotyping.

Thus the project is organized into a gathering of buildings sectioned in rhythm with the urban design. A united front is presented from the ground. One section settles there, bordering Marguerite Legros street, which in turn crosses the existing block, where the street makes two building façades face each other between two breaks.

这个项目建立在一系列的运动和静止的结合之上，以此为框架，设计师试图将街道的城市规模与周围的环境相结合。考虑到场地的密度，这个项目保留了建筑与街道之间必要的界限，最终既营造了私人住宅的氛围，又保有每个单元的独立性，与标准的混凝土式住宅群不同，脱离了那种刻板之感。

该项目被构建成城市环境下韵律性分段建筑的集合。从地面看来，建筑的前立面是统一的。坐落在半尺街的建筑与穿过现有建筑的勒格罗街相邻，在这两段路之间，两种建筑立面遥相对应。

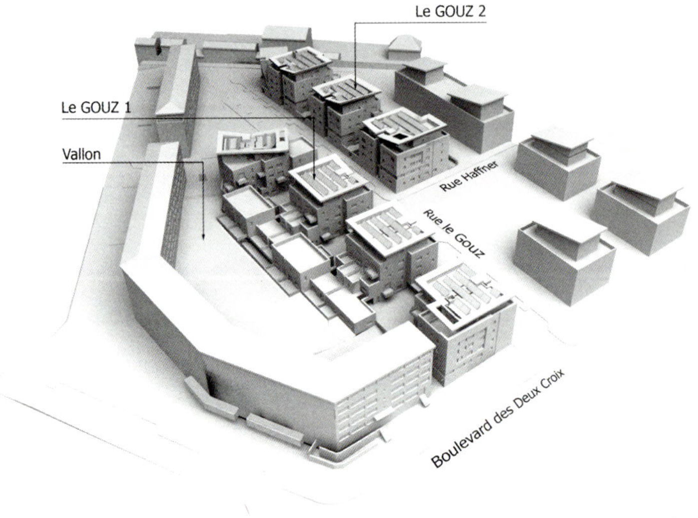

Thumbnail Aerial View 缩略鸟瞰图

FEATURE 特点分析

Shape

A well-handled gap separates the structures so that their individuality is maintained, yet they all have corresponding foundations so that a single and continuous reading of the urban landscape is achieved. Transparencies and visual escapes produced by the gap between structures find focus in the patios distributed among the large street-facing apartments. These patios initiate a smooth transition between public and private areas. In slight contrast to the regular rhythm of the buildings, a dynamic ensemble of rooftops tipped with slightly asymmetrical angles house attics for the apartments. The project flirts with false symmetry and vibrations established by slight variations of a serial logic. Moreover, the urban dimension earns its place amongst the strong goals of the urban project in Angers.

造型

以同一片地基为基础，该建筑巧妙地用间隔分割结构来保持各部分的独立性，因此实现了城市景观的连续性和单一性。建筑间的空隙产生的透明感和视觉差，使人们将焦点放在分布于临街公寓之间的露台中，实现了公共区和私人区之间的平稳过渡。略不对称的倾斜阁楼屋顶与这些建筑规则的韵律形成轻微对比，不完全对称与序列逻辑微变的摇摆外形，让这栋建筑保持着一种活泼的风格。此外，它的设计密度完全符合了昂热城市项目的目标。

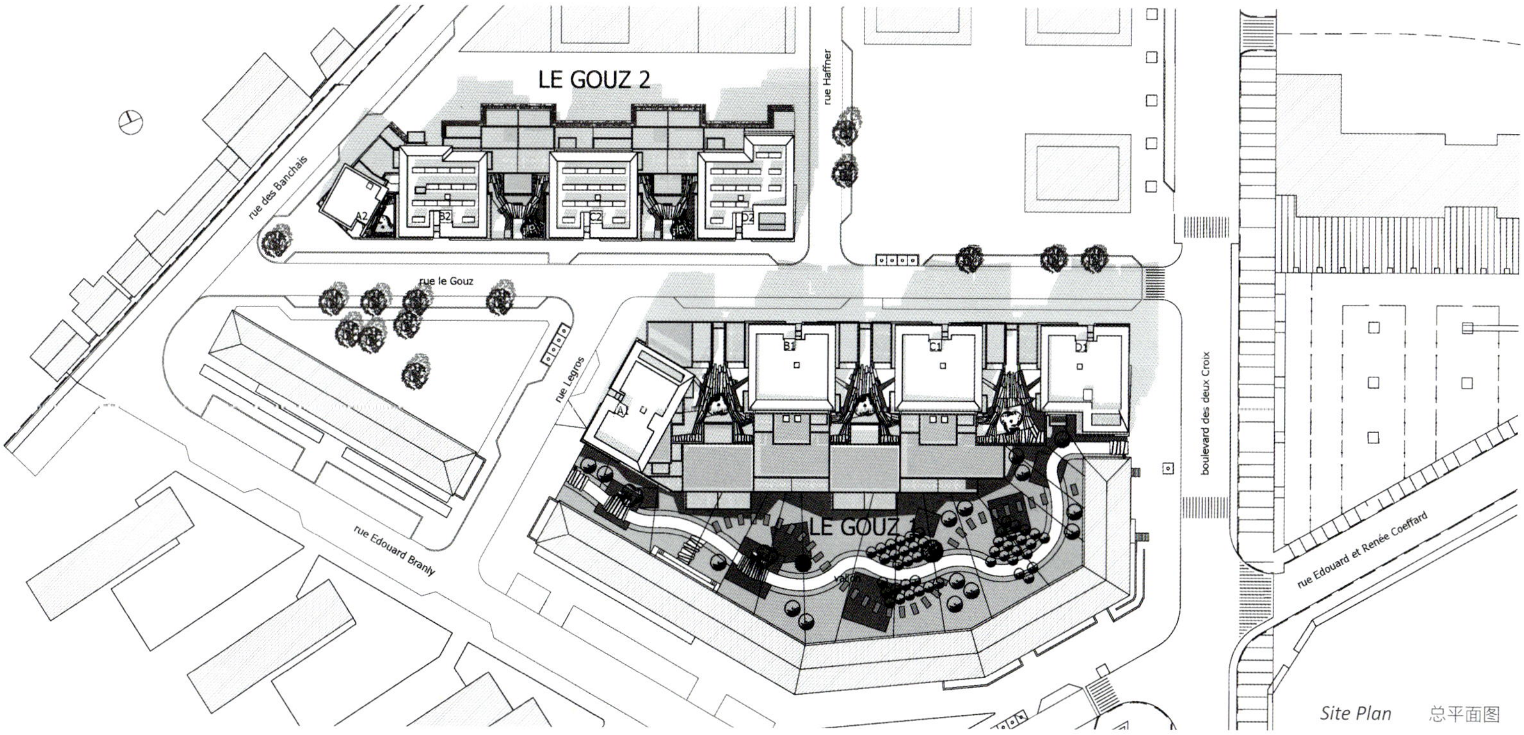

Site Plan 总平面图

Student Housing Duwo Delft, The Netherlands

Location: Delft,The Netherlands
Architect: Mecanoo Architecten
Gross Floor Area: 9,500 m^2
Number of Floors: 6
Photographer: Christian Richters

荷兰代夫特市 Duwo 学生公寓

地点：荷兰代夫特
设计公司：梅卡诺建筑事务所
建筑面积：9 500 m^2
层数：6
摄影师：克里斯汀・里克特斯

Surmount 突破点

ECONOMIC 经济
FACADE 立面
COLOR 色彩
MODULAR 模块化

The numbers of students enrolled at the Technical University of Delft has increased dramatically in recent years. This has created a sharp student housing shortage on the TU Campus. To tackle this problem, more than 1,700 homes will be built in the coming years. Because these homes are not yet readily available, the student housing foundation, DUWO, asked Mecanoo to design housing, which can be easily achieved, for the increasing group of foreign students.

In response, Mecanoo designed 3 towers which can be assembled with a high degree of prefabrication on the Leeghwaterstraat; sketch to completion, including the authorization process, is only one year. Through Mecanoo's design, the TU campus is able to achieve a specially designed, high quality residential housing complex in a very short period of time.

The 3 omni-sided residential towers have a total of 186 residential units, each 6 storeys high bordering a pasture. 3 façades are finished in dark masonry with squeezed joints and aluminium framed sliding doors. Large windows offer panoramic views of the surroundings. In the fall, when the climbers lose their leaves, the plant motif and green colour of the composite plates become visible. The 3 buildings are enveloped in a grass carpet strewn with natural and flagstones.

So that the project could be designed and realized within a year, a modular construction system was implemented, where each dwelling unit is completely pre-fabricated in the factory. The towers have a concrete core with staircase and elevator. A steel frame is mounted against the housing units like drawers in a cabinet loaded with a great variety of housing types. In the residences, students live on their own or with one other person. There are also small student houses with 3 rooms surrounding a common living room with a kitchen. On the ground floor of each tower are common facilities such as entrance, bicycle and storage space.

近几年来，到代夫特技术大学求学的学生数量急剧增加。因此该校面临宿舍楼紧缺的窘境。为解决这一问题，接下来的几年，学校将新建 1 700 户楼房。如此庞大的工程似乎难以在短期内实现，而为了解决学生住宿需求的燃眉之急，该校学生住宅委员会 DUWO 委托梅卡诺建筑事务所设计出建设周期短、易实现的住宅。

为实现目标，梅卡诺建筑事务所在 Leeghwaterstraat 大街设计了 3 栋公寓楼，都是可组装的、高级预置的结构，从设计到完成，包括验收，只用了 1 年的时间。此设计方案成功使校方在很短的时间内获得了 1 栋设计独特而高质量的住宅楼。

这 3 栋宿舍楼都是 6 层的设计，总共拥有 186 个住宅单元。立面全方位装饰过，每栋拥有 1 个绿色立面，其余 3 个立面则是紧密排列的黑色砌石连接铝框滑动窗的设计。宽大的窗户可以让人在此饱览周围的全景。而到了秋季，当藤蔓的叶枯萎时，复合镀层上植物的轮廓和镀层上的绿色变得清晰可见。值得一提的是，3 栋楼都被草皮包裹，如地毯一般延伸到石板路，突显绿色自然。

这个项目的设计能够在 1 年里实现的秘诀在于其模块化结构的设计，即每个住宅单元是在工厂预先定制好，再组装在主体楼上的。主体楼有 1 个包含了楼梯与电梯的混凝土芯。钢结构架起户型多样的住宅单元，好比安装在柜子上的抽屉。在这些单元里，有的是单人间，有的是双人间，当然还有些是 3 室 1 厅 1 厨的布局。每栋楼的首层入口处配备了基础公共设施，如入口、自行车停放处和储物空间。

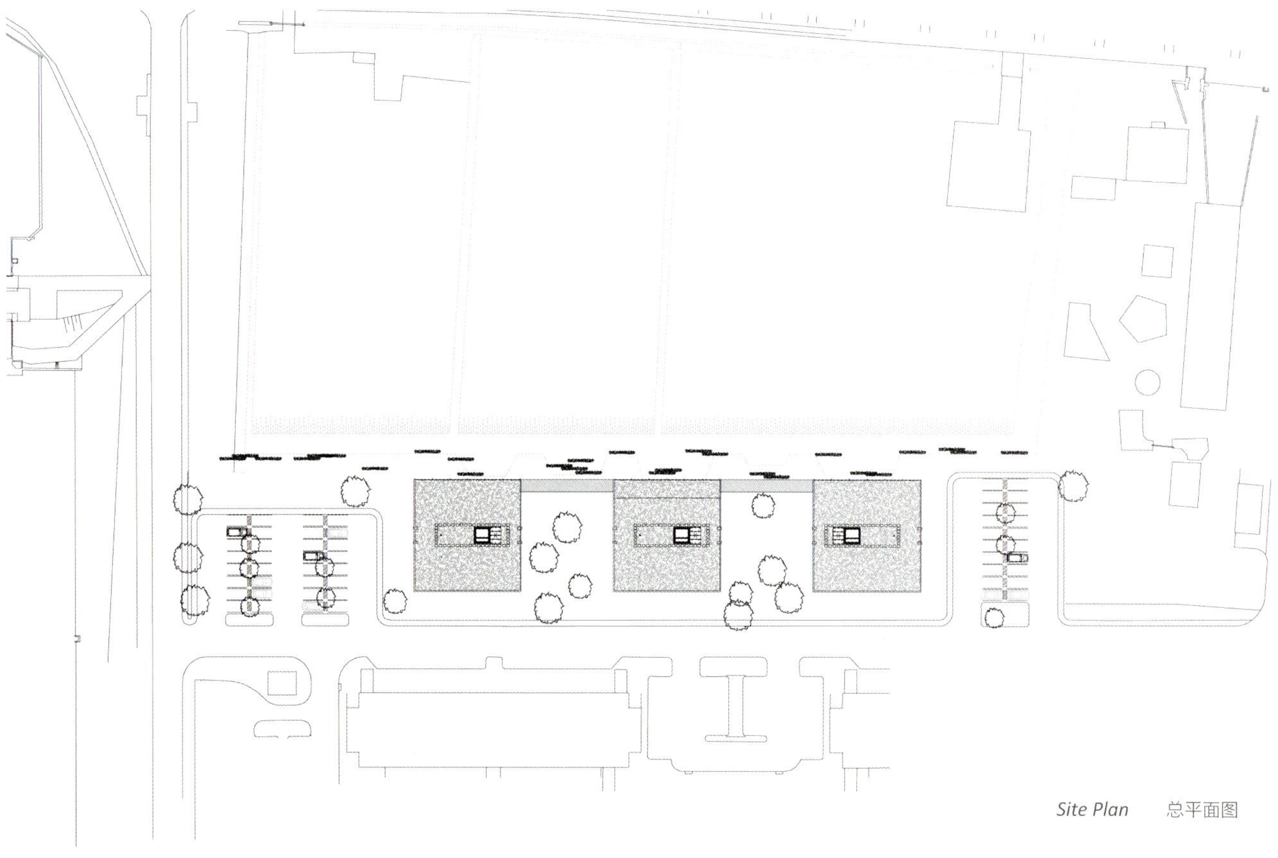

Site Plan 总平面图

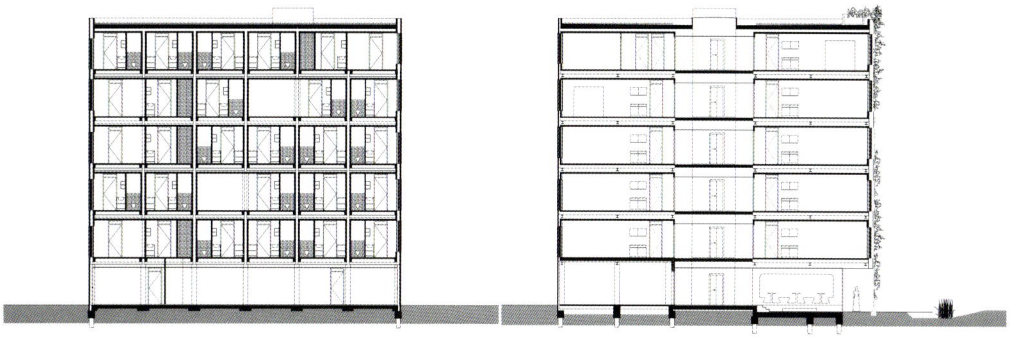

Section Plan 剖面图

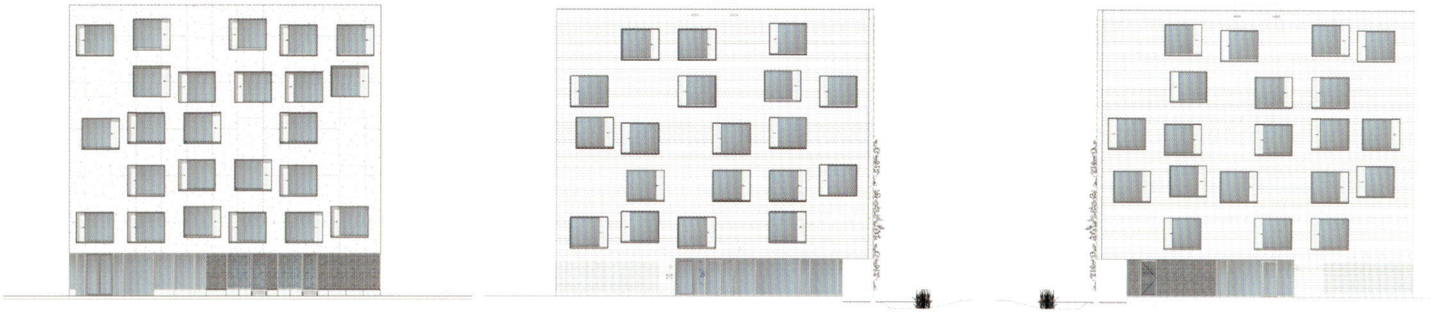

Elevation Plan 立面图

FEATURE 特点分析

Façade

Their playful composition creates an interesting picture not only on the exterior, but also on the interior: each room has a window in a different spot. On the pasture side, the three buildings display a layered green façade of composite plates in combination with a tubular frame that allows climbing plants to grow in the spring.

立面

窗口位置错落有致，不但营造出宿舍楼外立面的活泼感，而且使室内空间变得有趣；每个房间的窗户所处位置都不一样。而绿色立面则由复合镀层与管式桁架构成，这样设计便于在此空间内的藤蔓植物在春日里生长。

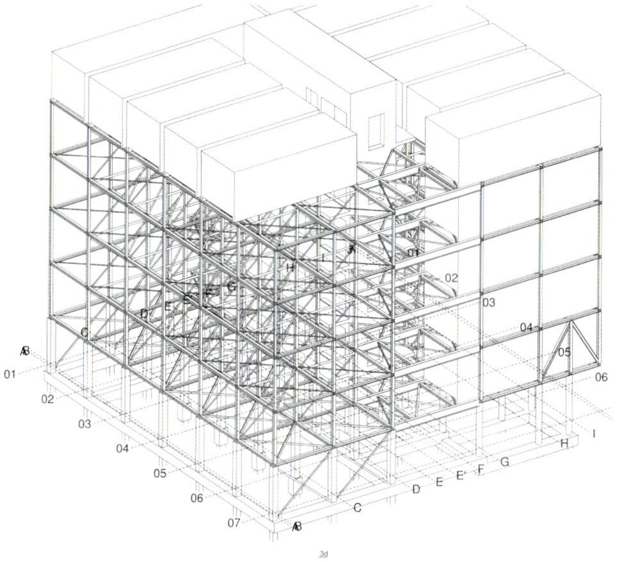

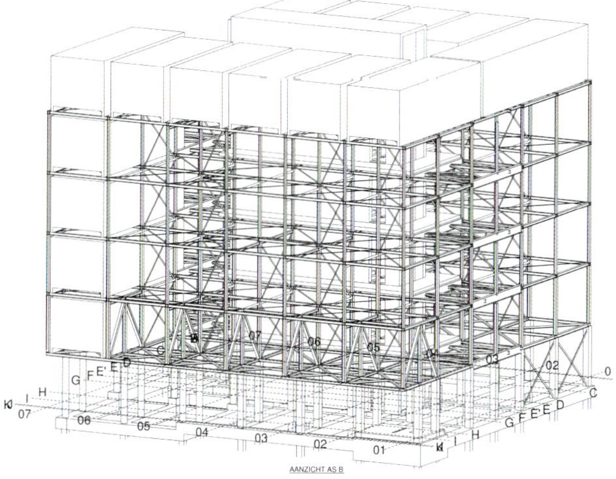

3D Structure Drawings 三维结构图

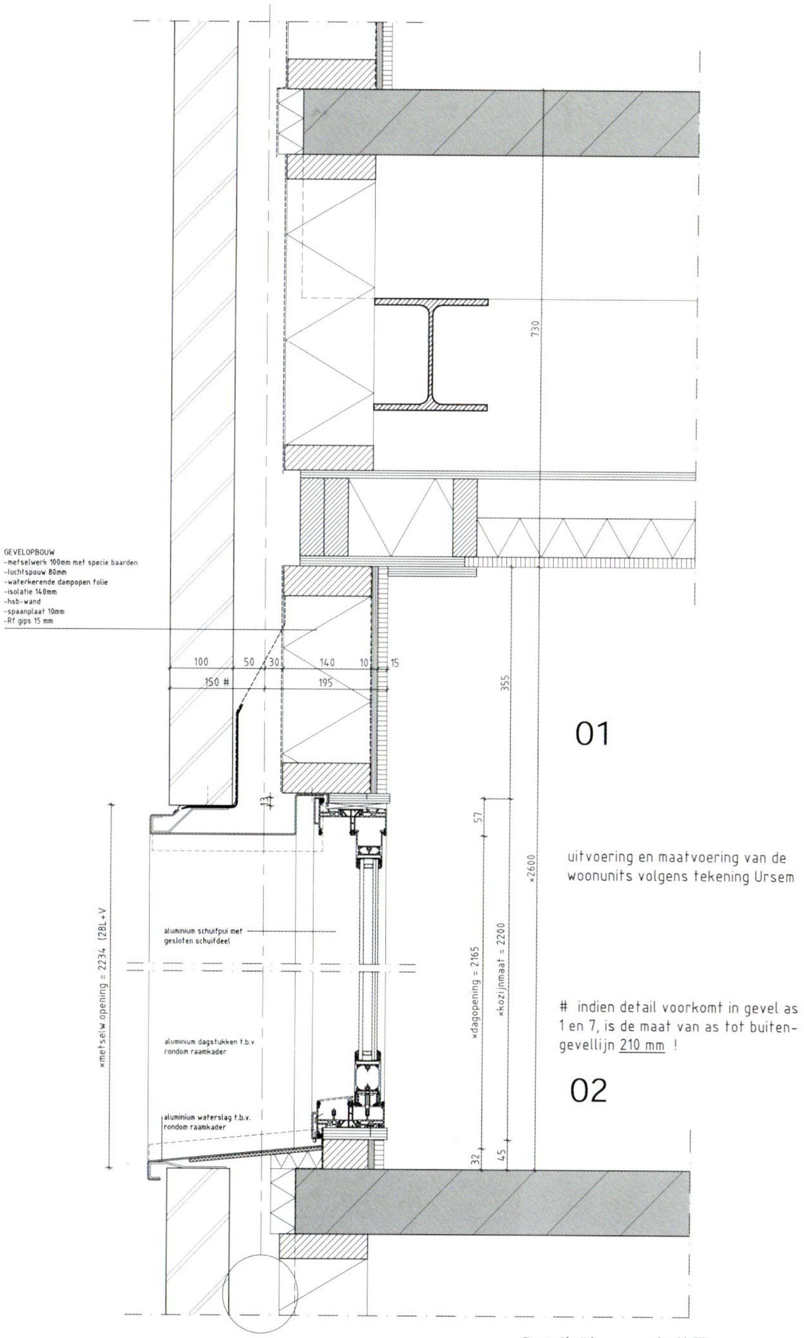

Detail Plan 细节图

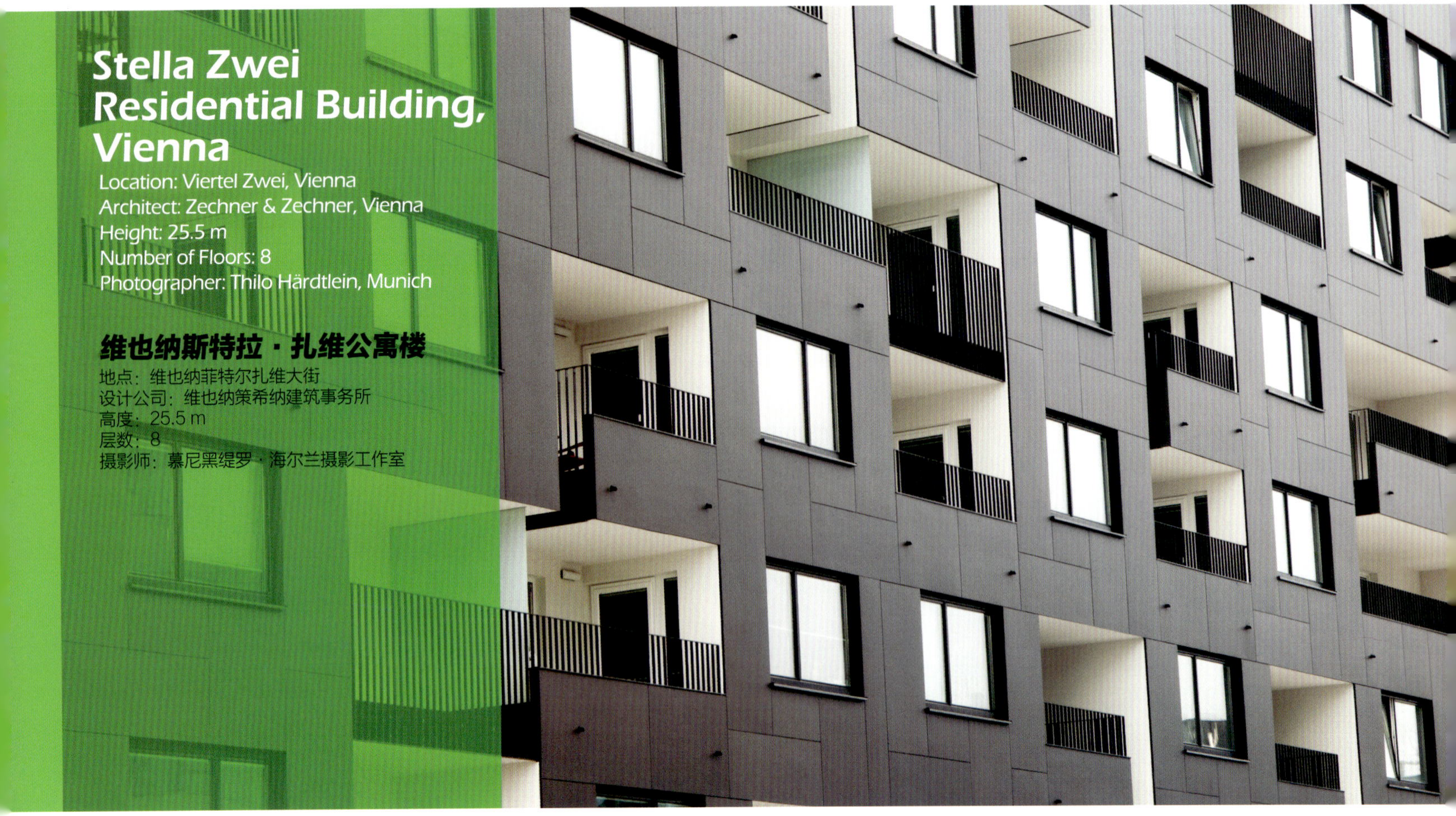

Stella Zwei Residential Building, Vienna

Location: Viertel Zwei, Vienna
Architect: Zechner & Zechner, Vienna
Height: 25.5 m
Number of Floors: 8
Photographer: Thilo Härdtlein, Munich

维也纳斯特拉·扎维公寓楼

地点：维也纳菲特尔扎维大街
设计公司：维也纳策希纳建筑事务所
高度：25.5 m
层数：8
摄影师：慕尼黑缇罗·海尔兰摄影工作室

Surmount 突破点

SKIN 表皮

FACADE 立面

BALCONY 阳台

ENERGY SAVING 节能

The newly arisen Viertel Zwei is characterised by the all-pervading attempt to develop an urban situation that generates a spatial definition for contemporary worlds of work. This Viertel Zwei, which is dedicated to the use of work in the tertiary and quaternary sector, service and information, has been configured as a self-contained place. The area is noted for having superb internal connection qualities and for the park-style quality of its public zones. The two elements of meadow and water prove that the necessary period of relaxation at work, the breaktime, has also been thought about in a visually present way and as an opportunity to feel. The only two exceptions from the application of work, which constitutes Viertel Zwei, are a hotel and a residential building.

The atmospheric qualities of the landscape planning, the use of water and green space, are as good living as they are for working. The use of work is also present in the residential building Stella Zwei. The ground floor zone is formed as a transparent zone and is used for offices. All of the remaining 6 floors and the roof level of the low energy building are occupied by apartments of 4 different sizes, from 50 m^2, 70 m^2 and 90 m^2 to the 155 m^2 penthouses at the top.

Stella Zwei is located in Viertel Zwei in Vienna's 2nd district – near the centre of town in attractive metropolitan surroundings. Alongside great infrastructure and good transport connections, Stella Zwei offers a range of open air spaces.

The 5,000 m² lake just around the corner and expansive green spaces in the nearby Prater district take the stress out of living in the city, and the whole area is laid out to be car free, with all parking below ground.

The building layout takes its design cue from the example set by Viertel Zwei. The eight-storey residential building, with its compact bars, extends the structure along the metro tracks, as a contrast to the flowing forms along Trabrennstraße and Trabrennbahn, and creates a border to Viertel Zwei in the direction of Stella-Klein-Löw-Weg.

The external form of the building is defined by the non-uniform placement of loggias, some of which are planned to span two stories.

Consistently executed reflections of the floor plans, which result in changes in the positioning of the loggias, creates an outward illustration of these contrasts on the façade. The façade can be read as a picture of the spatial configurations behind. The façade is in dark-grey eternit, the air spaces and the loggias in white render. In this way, one's own apartment is understood as an individualised element within a system of apartments.

The ground floor design is an inset base that is transparent towards the lake. Along with shared facilities like the waste disposal room, and storage rooms for bikes and children's buggies, there are offices with modern equipment. A glass lobby that traverses the entire floor provides open vistas and access to the 7 residential floors.

The top floor is recessed, and will be covered with bright aluminium panels, as an interesting contrast to the other floors. This step back reduces the apparent height of the building and creates large terraces for the penthouse apartments.

The 72 apartments, which range in size from 49 m² to 115 m², have sophisticated and functional floor plans, generous living rooms, all have their own loggia, fully equipped kitchens and bathrooms as well as desirable wooden floors.

There are 6 penthouses on the top floor, ranging in size from 98 m² to 156 m², which are provided with generous terraces.

新建成的菲特尔扎维大街用全方位的特点来展示城市的概念，为当今世界的工作诠释了空间的定义。菲特尔扎维大街致力于工作上的三级和四级机构：服务部和信息部，并以此形成一个自成一体的机构。该区域还配有高级内部联系服务和公园式的公共区。大街还设有草坪和水景，使人们在工作期间不忘休息，在放松身心的同时既能在视觉上观赏到草坪和流水，还能抓住难得的感受自然的大好时机。在菲特尔扎维大街的建筑构成中，有两个例外，即它有酒店和公寓楼。

景观规划中对空气质量的考虑以及对流水和绿地的使用，使整个环境既益于生活又利于工作。用于办公的场所在斯特拉扎维公寓楼中也有所体现。首层是被设计成透明的区域，用作办公区。其余 6 层和顶层采用低能耗设计，有 4 种不同尺寸的房型：50 m²、70 m²、90 m²，以及顶楼的 155 m² 复式套房。

Sketch 概念草图

Façade Pattern 立面图案

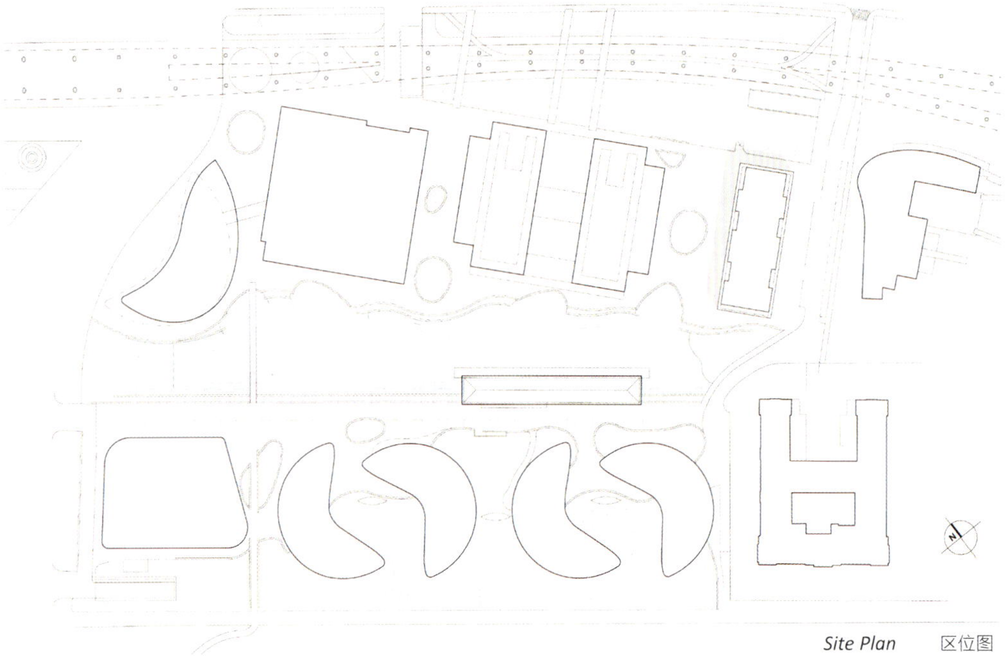

Site Plan 区位图

ANSICHT STELLA-KLEIN-LÖW-WEG

Elevation Plan 立面图

斯特拉扎维公寓楼位于维也纳第 2 区的菲特尔扎维大街，毗邻充满城市魅力的城镇中心。除了有大型的基础设施和良好的交通路线，斯特拉扎维公寓楼还提供了宽敞的露天空间。

5 000 m^2 的大湖就在拐角，加上普拉特区的开阔绿地，让城市生活的压力烟消云散。此外，整个区域的车都免费停放于地下。

建筑的格局设计灵感来源于菲特尔扎维大街的布局。八层高的公寓楼和附带的小巧酒吧，沿着地铁轨道延伸，与流线型的 Trabrennstraße and Trabrennbahn 形成对比，并在菲特尔扎维大街和 Stella-Klein-Löw-Weg 之间形成明显的分界线。

建筑外部的阳台采用错落的放置形式，其中有些还设计成跨越两层的结构。

为了与平面图的设计一致，阳台的位置发生了变换，这种变换反映在立面上，形成强烈的对比。立面可以被理解成空间配置而成的一幅画面。立面采用的是深灰色的石棉水泥，空间和阳台则采用白色。如此一来，个人公寓就可被理解为一个公寓系统里的个体元素。

首层设计为朝湖的透明嵌入式空间，里面包含多种公共设施，如废品储放室、自行车和儿童手推车等的贮藏室，以及其他带现代化设备的办公室。玻璃式的大堂穿过整个楼层，不仅是上面 7 层的住户的进出口，还创造了一片宽敞的远景。

顶层往里凹进，并覆盖了一层铝合金面板，这和其他楼层形成了有趣的对比。这样的设计感觉明显降低了建筑的高度并为复式套房铺设了宽敞的露台。

公寓楼共有 72 间公寓，尺寸从 49 m^2 到 115 m^2 不等，均配有全面而多功能的设计，如特大起居室、独立阳台、设备齐全的厨房和浴室，以及令人向往的木地板。

顶层有 6 间复式套房，尺寸从 98 m^2 到 156 m^2 不等，皆配有宽敞的露台。

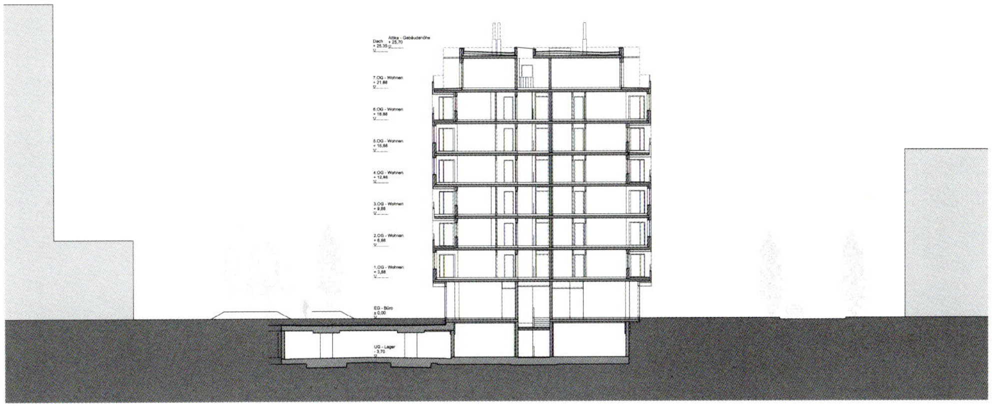

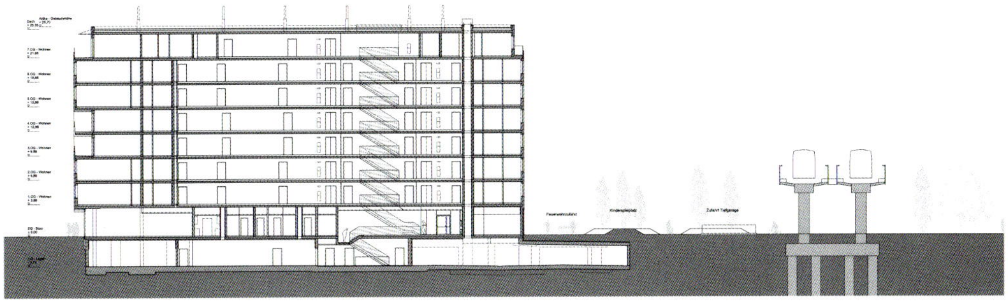

Section Plan 剖面图

FEATURE 特点分析

Façade

The loggias are bright spaces cut into the anthracite grey box, and this creates a playful and varied façade. The entire façade is a game between dark and light, inside and outside. This creates an inverted black and white picture, which increases the affinity of the occupants for their living space.

立面

阳台就像是切入灰色无烟煤方块体量的白色空间，为建筑创造了有趣多变的立面。整个立面是一场光与暗、内与外之间的游戏，这种内外之间黑白倒转的画面，让住户对他们的居住环境产生一种亲切感。

STELLA
ZWEI

Floor Plan 楼层平面图

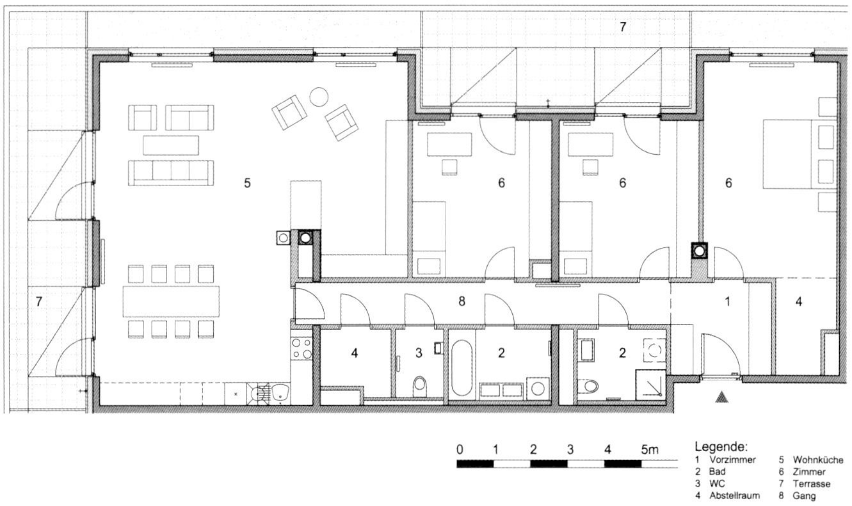

Floor Plan XL 超大户型图

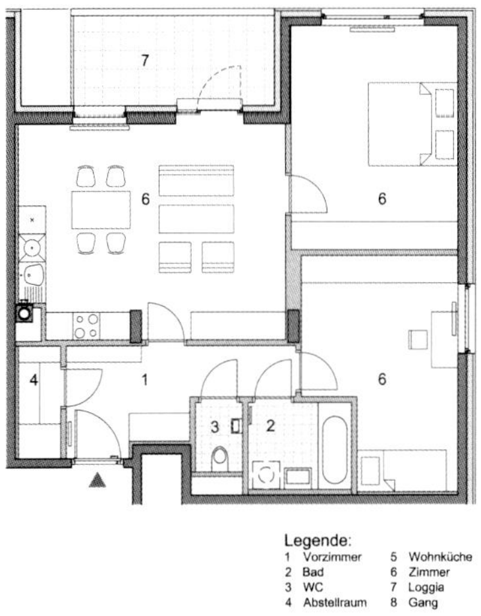

Floor Plan M 中户型图

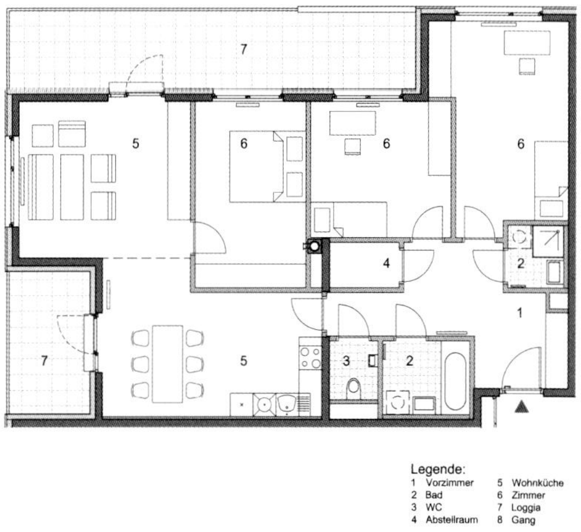

Floor Plan L 大户型图

Iceberg in the Aarhus Docklands, Denmark

Location: Aarhus Docklands development, Denmark
Architect: Julien De Smedt Architects
Site Area: 25,000 m^2
Number of Floors: 10

丹麦奥尔胡斯港区冰山住宅综合楼

地点：丹麦新奥尔胡斯港区
设计公司：JDS 建筑事务所
占地面积：25 000 m^2
层数：10

Surmount 突破点

ECONOMIC 经济
SHAPE 造型
DAYLIGHTING 采光
VIEW 视野

JDS and Aarhus-based CEBRA, in collaboration with the Dutch firm SeARCH and French architect Louis Paillard wins the competition to build a 25,000 m^2 housing complex in the new Aarhus Docklands development. The project will be developed by the Tækker Group and Brabrand Boligforening with an expected completion date of spring 2010.

The Aarhus Harbour development provides a huge opportunity for Denmark's second largest city to develop in a socially sustainable way by renovating its old, out-of-use container terminal. The area is meant to become a living city quarter, comprised of a multitude of cultural and social activities, a generous amount of workplaces, and of course, a highly mixed and diverse array of housing types.

The Iceberg Project seeks to locate itself within the goals of the overall city development. A third of the project's 200 apartments will be set aside as affordable rental housing, aimed at integrating a diverse social profile into the new neighborhood development.

"With the Iceberg we get unique housing qualities as well as a city architectural expression of the highest quality," says Kent Martinussen, administration director of DAC (Danish Architecture Centre).

"Aarhus will get a fantastic harbour front with unique architectural buildings that both in appearance and functionality prove that we are a city of grand ambitions. Our desire for this area goes beyond just a façade without life and purpose. We want a living city where everybody thrives, bth those who live and those who work in this 'City near the harbour / De Bynære Havnearealer'. Projects of this calibre are a big step towards this goal," said Mayor, Nicolai Wammen.

The Iceberg was mainly inspired by the amazing location. The architects decided to break up the building into several parts and levels in order to create optimal ocean views for as many apartments as possible. Therefore, the building was split up in a criss-cross manner – inspired by floating icebergs constantly breaking away.

In order to prioritise views and natural light even further, the architects developed sloping roof tops providing stunning views in several directions. The roofs rise and fall in peaks and canyons and are visually displaced from each other just like the motions of real icebergs. Creating a logical harmony with the heaven-aspiring triangular spires, the façade was split into loosely combined triangles that lend the façade a dynamic feel. This provides a harmonious connection between the complex as a whole and its details. The large, white triangles and the flickering surface gave rise to the name of the complex: The Iceberg.

The Iceberg breaks with traditional architecture typically involving a "wall" of homes. The complex takes into consideration those homes located at the back by creating peaks and canyons that ensure optimal light conditions and ocean view.

Roof Floor Plan 屋顶平面图

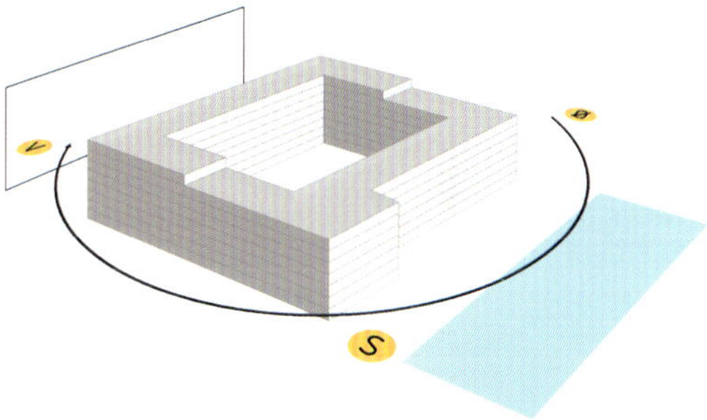

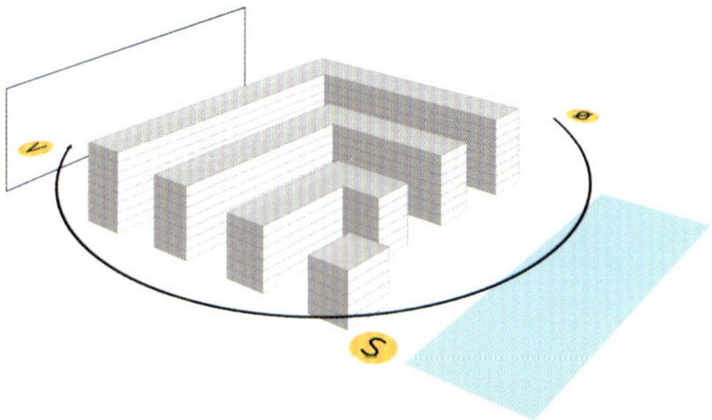

Views Analysis Plan 视线分析图

Light Analysis Plan 日照分析图

Section Plan A-A　剖面图 A–A

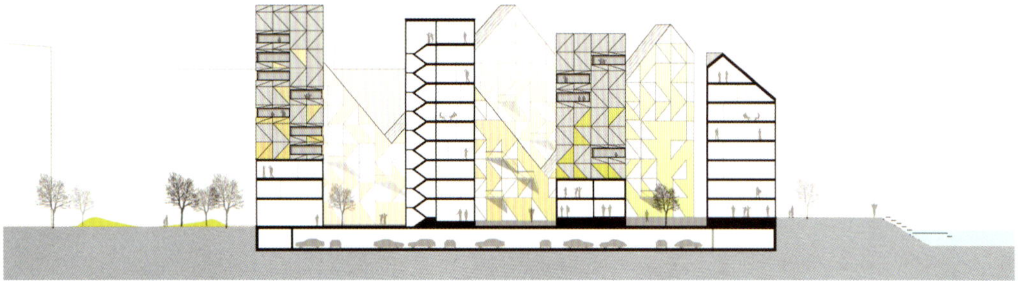

Section Plan B-B　剖面图 B–B

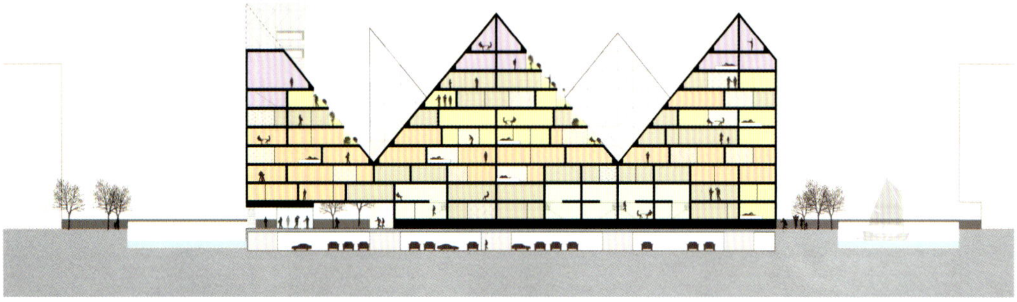

Section Plan C-C　剖面图 C–C

Elevation Plan 1　立面图 1

Elevation Plan 2　立面图 2

丹麦的 JDS 建筑事务所、总部在奥尔胡斯的 CEBRA、荷兰的 SeARCH 公司及法国的建筑师 Louis Paillard 一起在竞争中取胜，共同设计新奥尔胡斯港区发展项目中占地 25 000 m^2 的住宅综合楼。该项目由 Tække r集团和 Brabrand Boligforening 公司开发，在 2010 年春天完工。

通过重整其老旧、过时的集装箱码头，奥尔胡斯港区项目正为这个丹麦第二大城市的社会可持续发展之路提供了巨大的机遇。港区计划改造成一个充满活力的城市一角，其中包括众多的文化和社会活动空间和广阔的工作场所，当然还有高度综合的一系列多样化住宅类型。

冰山项目组试图在城市总体发展目标中寻找自己的定位。项目总共包括 200 间公寓，而其中的三分之一将被定位为公众可负担得起的出租房，使一个多样化的社会融入到这个新社区的发展中。

“多亏了冰山项目组，我们得到了独特的住宅品质，同时还有城市建筑最高品质的表达。”丹麦建筑中心的行政总监 Kent Martinussen 说道。

“奥尔胡斯将会成为一个梦幻的海港，它前面的这栋独特的建筑无论在外观上还是功能上都证明我们是一座富有豪情壮志的城市。我们的愿望不是一栋没有生命和目标的建筑，我们想要的是一座朝气蓬勃的城市，使所有在这个港口城市居住或工作的人都感到生机盎然。而拥有如此高品质的这个项目已经朝该目标迈了一大步。”市长 Nicolai Wammen 说道。

冰山的灵感来源于其令人惊叹的地理位置。设计师决定把建筑分为几个部分和层次，努力让最多的房间欣赏到美丽的海景。因此，受漂浮的冰山经常分裂这种现象所启发，建筑被分裂成纵横交错的形式。

为了使视野更宽广，自然采光更充足，设计师采用了倾斜的屋顶，以保证不同方向的开阔视野。屋顶的上升和下降形成了“山峰”和“峡谷”，看起来就像真正的冰山一样。高耸入云的三角尖顶一座连着一座，把立面变成了松散结合的三角形，看起来更具动感，把建筑的整体和细节都和谐地融合在一起。而这些大型的白色三角形和闪耀的立面就是建筑名字的由来——冰山住宅综合楼。

冰山住宅综合楼打破了传统建筑中“墙”的概念。考虑到背面房间的视野，设计师采用“山峰”和“峡谷”式的造型以确保其良好的采光条件和海景视野。

FEATURE 特点分析

Shape

The project's main obstacle is the density set up for the development; the desired square meters are in conflict with the specified site height restrictions and the overall intentions of providing ocean views along with good daylight conditions. The Iceberg negotiates this problematic, by remaining far below the maximum heights at points and emerging above the dotted line at other moments. "Peaks" and "canyons" form, eliciting the project's iconic strength while ensuring that all flats will be supplied with a generous amount of natural lighting and waterfront views.

造型

项目的主要障碍是开发区计划的密度。所需的面积与指定场地的高度限制，以及想拥有海景和良好日照条件的整体意愿相冲突。冰山项目组对此进行了协商，决定最高的地方只保留点，其他地方以虚线连接以形成“峰”和“谷”，突出项目的标志性实力，同时确保所有房间都能享受到充足的自然光和海景。

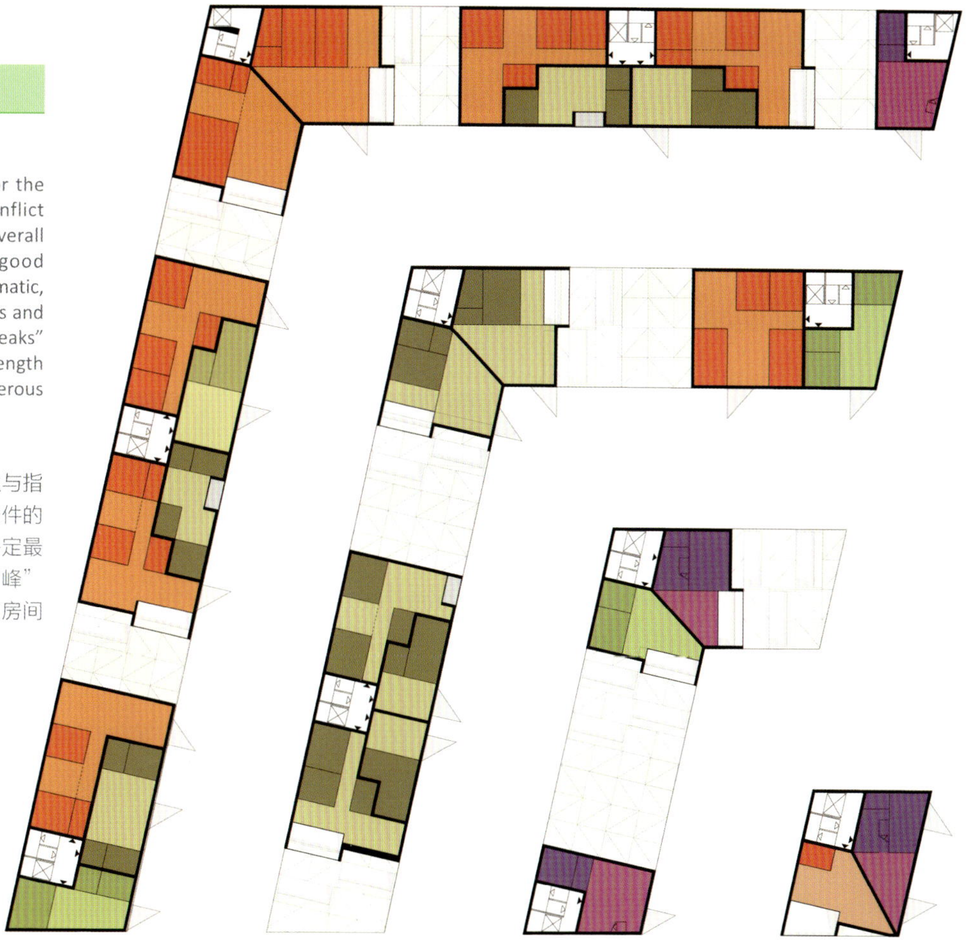

4th Floor Plan 4层平面图

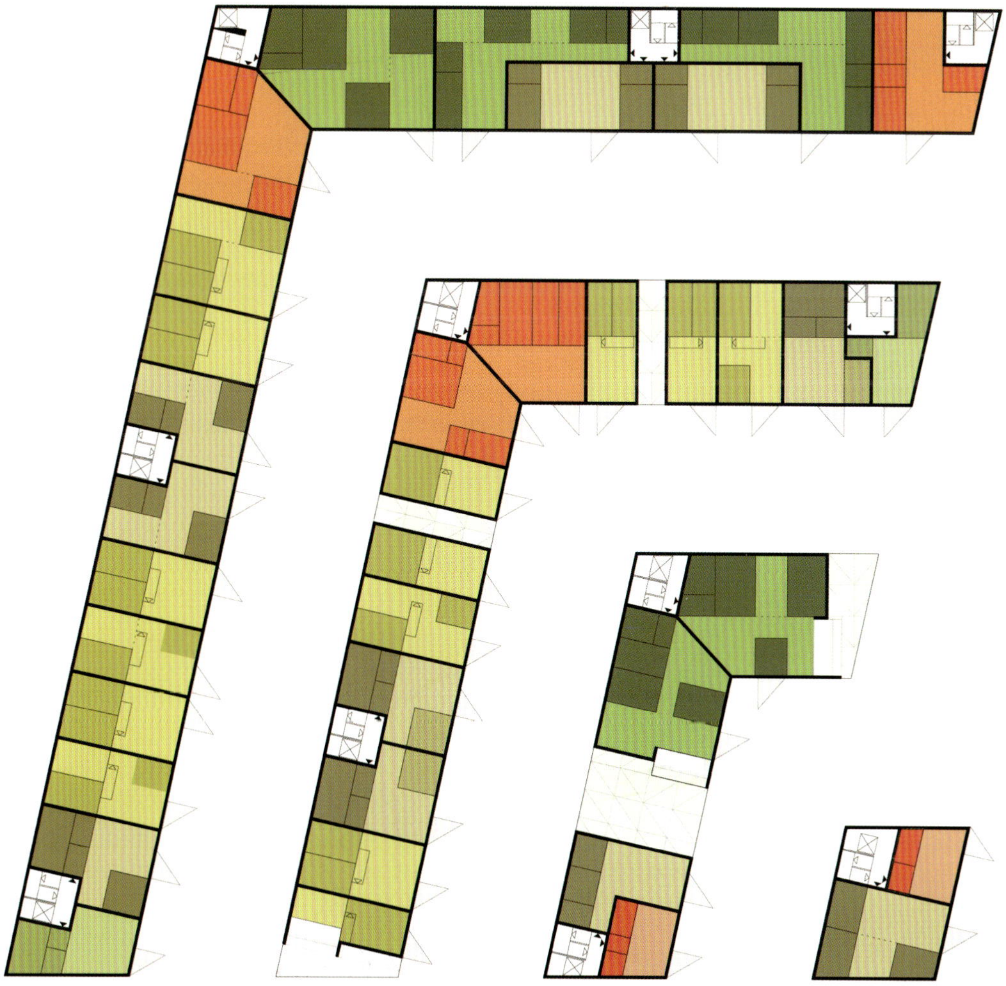

2nd Floor Plan 2层平面图

The Wave in Vejle, Denmark

Location: Vejle, Denmark
Architect: Henning Larsen Architects
Gross Floor Area: 14,000 m²
Number of Floors: 9
Photographer: Thomas Molvig

丹麦瓦埃勒市的波浪住宅楼

地点：丹麦瓦埃勒市
设计公司：亨宁·拉尔森建筑事务所
建筑面积：14 000 m²
层数：9
摄影师：托马斯·穆维奇

Surmount 突破点

VIEW 视线

SHAPE 造型

BALCONY 阳台

ENVIRONMENT 环境

The new landmark of Vejle is a distinctive residential building of 9 storeys with a magnificent location by the bay Skyttehusbugten and Vejle Fjord.

Architecturally, the building relates to its fantastic location by Vejle Fjord. In daytime, the building will be characterised by the soft movements of the waves reflected in the water surface of the fjord. At night, the characteristic profile of the Wave will appear as an undulating mountain landscape of light and colour.

The building comprises 100 attractive flats - many of which are in two levels. The layout of the flats ensures a splendid view from all balconies. Together with the public promenade in front of the building, the Wave creates a beautiful connection between the fjord, landscape and town and consolidates the different elements in a clear and recognisable signature.

波浪住宅楼位于丹麦 Skyttehusbugten 海湾和瓦埃勒市港湾沿岸，9 层高的独特外观使之成为该市的新标志性建筑风景。

建筑坐落在瓦埃勒市港湾最优越的位置上。白天，建筑楼轻柔的波浪外形将倒映在港湾的水面上，独特而醒目；到了晚上，这个波浪造型就会变成连绵起伏的山景，被一层柔光和斑斓色彩轻轻笼罩着，美轮美奂。

住宅楼内包含 100 个独具魅力的公寓，多为复式结构。公寓的布局确保了楼面上所有的阳台上都能观赏到壮阔的海湾美景。公寓楼前部的公共长廊和建筑楼独特的波浪造型把整个港湾、景观和小镇完美地联系在一起，使不同的元素和谐地融合成一个清晰可辨的景观标志。

Sketch 草图

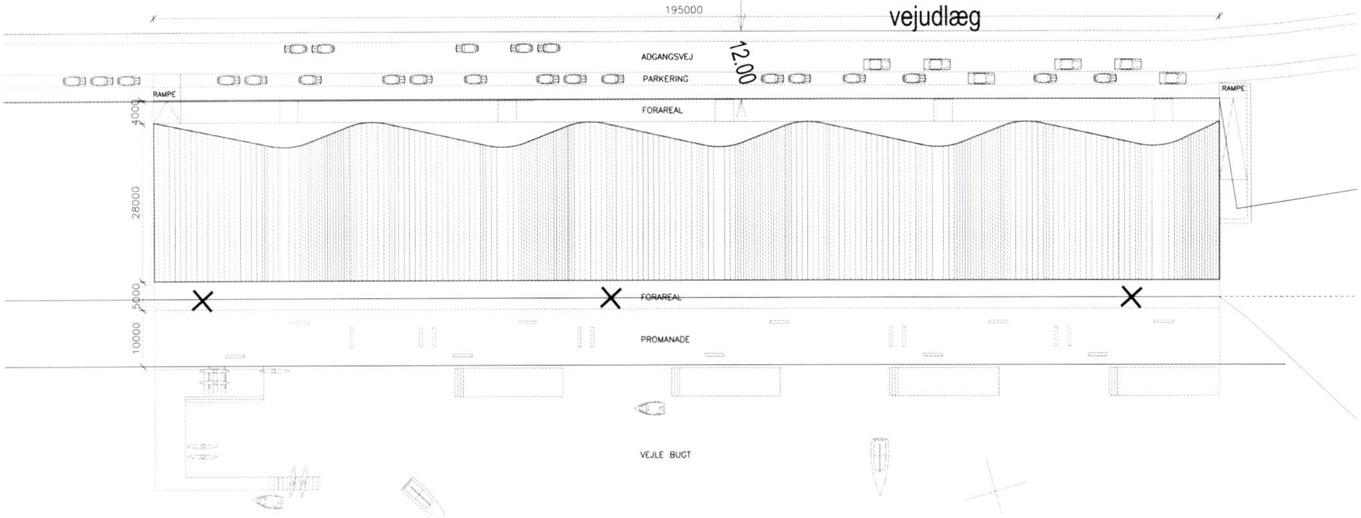
195000
vejudlæg
ADGANGSVEJ
12.00
PARKERING
RAMPE
FORAREAL
4000
28000
5000
FORAREAL
10000
PROMANADE
VEJLE BUGT

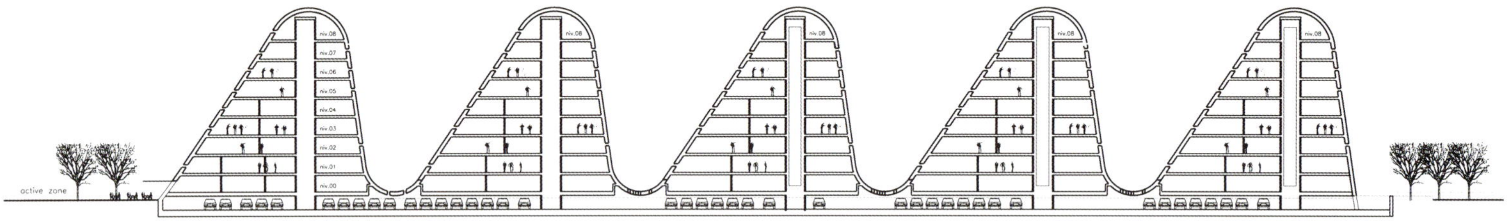

Longitudinal Section 纵向剖面图

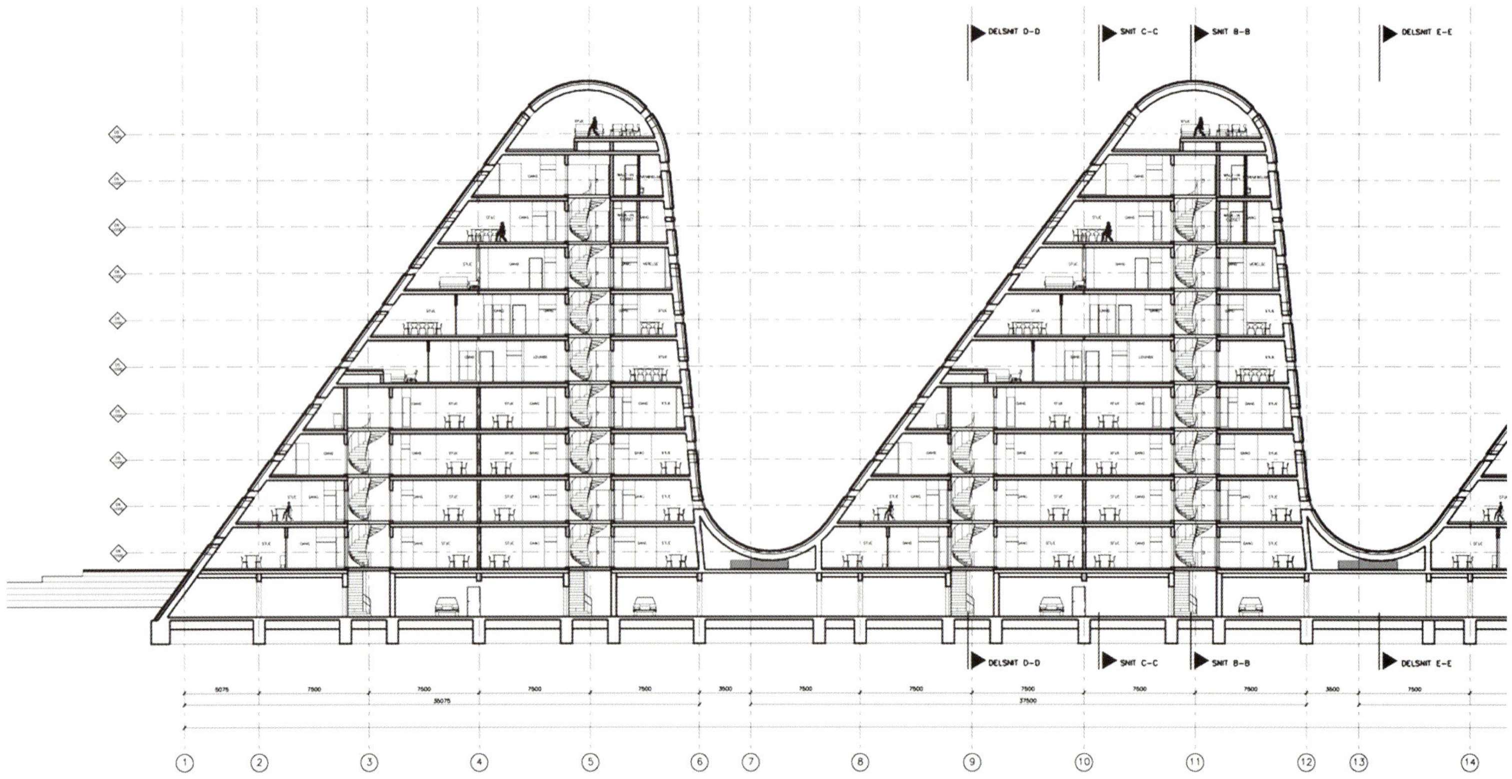

Section Detail 剖面细节图

FEATURE 特点分析

Shape

The wave shaped apartment fully uses and integrates into Vejle Fjord, both respecting and challenging its location.

造型

波浪式造型的住宅充分利用了港湾的地理环境，并与之完美融合，既尊重当地环境的风貌，又极富挑战性地将其运用到建筑设计中。

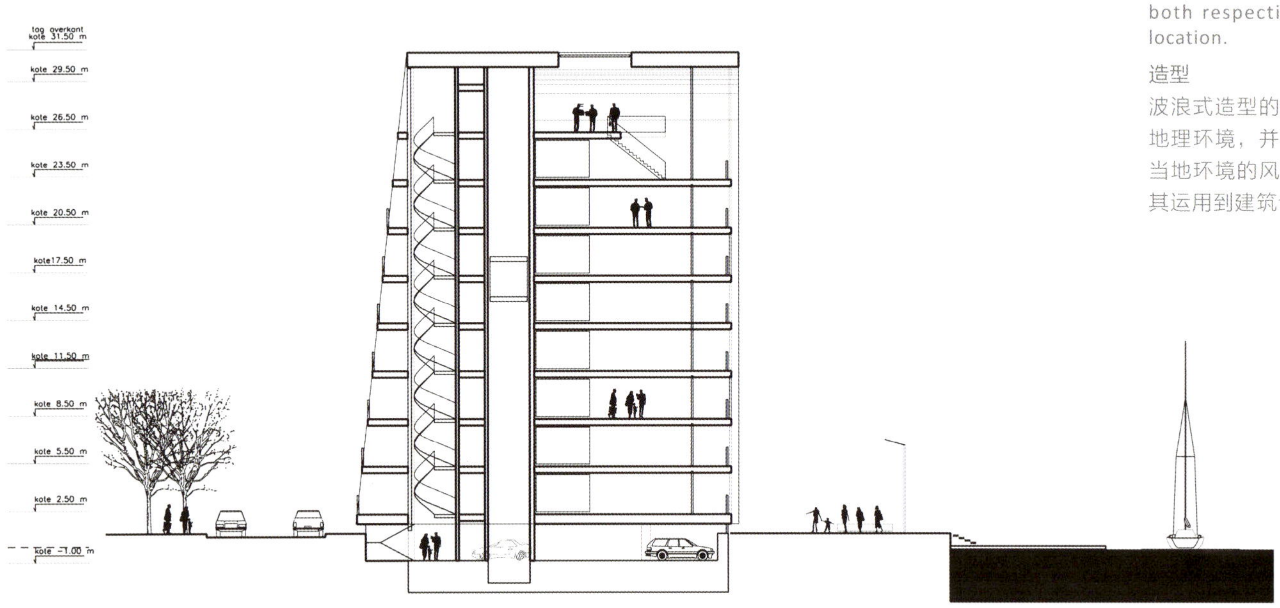

Transverse Section 横向剖面图

Calista Luxury Resort, Antalya-Belek, Turkey

Location: Antalya-Belek, Turkey
Architect: gmp von Gerkan, Marg and Partners Architects
Site Area: 120,000 m²
Gross Floor Area: 75,745 m²
Number of Floors: 6
Photographer: Marcus Bredt & Özkar Sanayi ve Ticaret A.S.

土耳其安塔利亚市卡利斯塔豪华度假村

地点：土耳其安塔利亚市贝莱克海滨
设计公司：冯・格康、玛格及合伙人建筑师事务所
占地面积：120 000 m²
建筑面积：75 745 m²
层数：6
摄影师：Marcus Bredt & Özkar Sanayi ve Ticaret A.S.

Surmount 突破点

VIEW 视野

SHAPE 造型

BALCONY 阳台

MATERIAL 材料

The Calista Luxury Resort is a retreat that is truly unique not only in Turkey but the whole world. This luxurious complex impresses with its unusual architecture and excellent service. The five-star hotel nestles idyllically amidst a pine forest covering a total surface area of 120,000 m² and an extraordinary outlook to the magnificent blue sea. In Calista Luxury Resort the variety of events and activities are endless and the entertainment and recreation we provide cater to every age, taste and culture. The Calista Luxury Resort is 4 km away from the centre of Belekand, overlooking the coast of Antalya and 35 km from Airport Antalya.

Situated right on its own private beach, this luxury hotel boasts fascinating architecture, paradisiacal grounds and a vast selection of restaurants and entertainment options to choose from. The Calista Luxury Resort looks as if a squadron of UFOs picked the most beautiful spot on Earth to land in. Around 20 cylinder-shaped buildings make up the 5 star luxury hotel and are spread out on.

The new hotel site fits discreetly into the surrounding countryside of pine groves and beaches. It combines both luxurious spaciousness with harmony and manageability of scale. Guests of the five-star hotel are accommodated in seven three to six-storey rotundas, each accessed via a central roofed atrium with galleries running round. On the outside, balconies provide an unobstructed view of the surrounding nature. For individualists, there are also 12 two-storey, likewise round villas available, with roof terraces and direct vistas of the sea. The presidential suite is on the top floor of the central building.

无论是在土耳其还是整个世界上，卡利斯塔豪华度假村都是一个真正的、独一无二的度假胜地，其非凡的建筑风格和卓越的服务给人留下了深刻的印象。这个五星级酒店占地 120 000 m^2，坐拥壮阔无边的蓝色海洋全景，四周还环绕着一片松树林，充满诗情画意。此外，卡利斯塔豪华度假村里的各种娱乐和活动层出不穷，使不同年龄层、不同品位、不同文化背景的人群都能在此尽兴。卡利斯塔豪华度假村离贝莱克海滨中心 4 km，离安塔利亚机场 35 km，可以俯瞰整个安塔利亚海岸。

卡利斯塔豪华度假村的酒店位于其私人海滩上，拥有迷人的建筑外观、天堂般的庭院，以及众多可供选择的餐厅和娱乐场所。卡利斯塔豪华度假村看起来就像是 UFO 群选择了这个地球上最美丽的地方作为憩息地。20 个圆筒形建筑一字排开，组成了这座五星级的豪华酒店。

这座新建成的酒店低调地融入到在松树林和海滩环绕的乡村环境中，糅合了豪华的空间、和谐的氛围和规模的管理。这座五星级酒店的客房都安排在 7 座 3~6 楼的圆形建筑中，客人可通过被画廊环绕的带屋顶的中庭进入各层。在酒店外部的阳台上，宾客可以对周围的景色一览无遗。对于喜欢享受个人空间的人群，酒店还配有 12 幢 2 层高的圆形别墅，并带有屋顶露台，可以直观海景。酒店中央的顶层是总统套房。

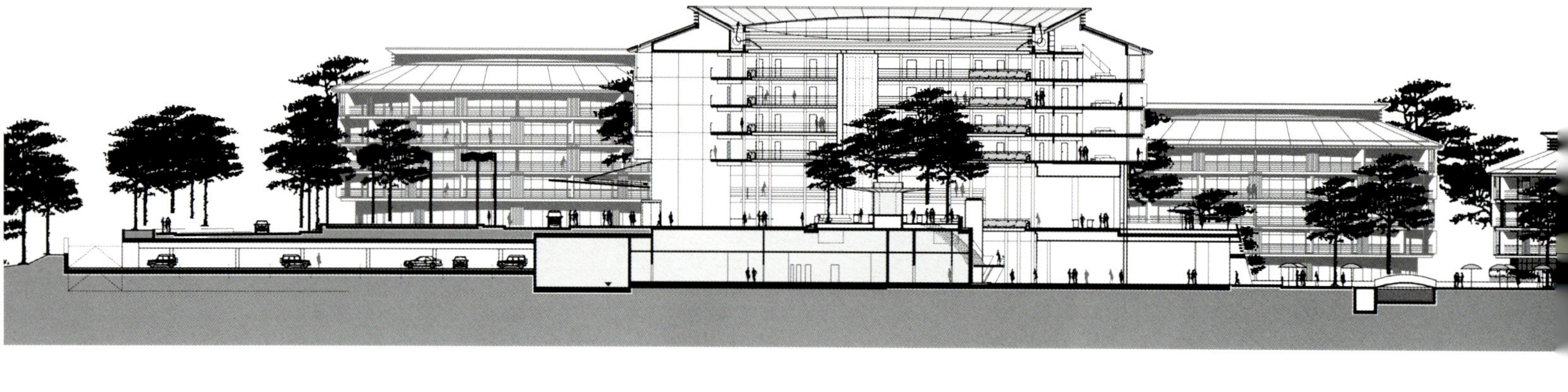

FEATURE 特点分析

Material

White concrete, light zinc roofs, glass façades, and the wood of the railings, blinds and pergolas match the typical building style of the region, giving the site a homogeneous appearance.

材料

白色混凝土、光锌屋顶、玻璃幕墙、木质栏杆、百叶窗，以及绿廊都与该地区的典型建筑风格相融合，同时也给予了场地一个同质的外观。

Hedorfs Residence Hall, Frederiksberg, Denmark

Location: Frederiksberg, Denmark
Architect: KHR arkitekter
Site Area: 1,700 m^2
Number of Floors: 4

丹麦菲特烈堡 Hedorfs 学生公寓

地点：丹麦菲特烈堡
设计公司：KHR 建筑设计公司
占地面积：1 700 m^2
层数：4

Surmount 突破点

WINDOW 窗
FACADE 立面
VIEW 视野
DAYLIGHTING 采光

Buildings made with sandwich panels have a reputation for being dull boxes. In Hedorf Residence Hall at Frederiksberg KHR arkitekter shows that buildings cast in concrete can be both creative and inspiring.

Inspired by the bay windows at traditional Frederiksberg apartments KHR arkitekter has designed a geometrically shaped bay window, which creates an expressive and dynamic façade. The bay windows are placed above the floor to give more spacious and dynamic apartments.

A small triangular window is placed in top corner of the bay window with the broad concrete wall behind. The result is a slim expression in the building because of the reflection from the window in the window ledge. The windows also give a kaleidoscopic effect where the expression of the building changes according to where you stand. From one side the façade opens up with windows, from the other side the closed bay windows with their organic movement characterize the façade. Thanks to the bay windows there is a play of colors between light and shadow all day independent of the light. And the wave motion on the façade makes the hard material appear soft and organic.

Hedorf Residence Hall is situated close to Flintholm Station, CBS and Niels Brock. The residence hall has 4 floors with a total of 52 one bed apartments. Each floor has a common kitchen where the students can meet, cook and socialize.

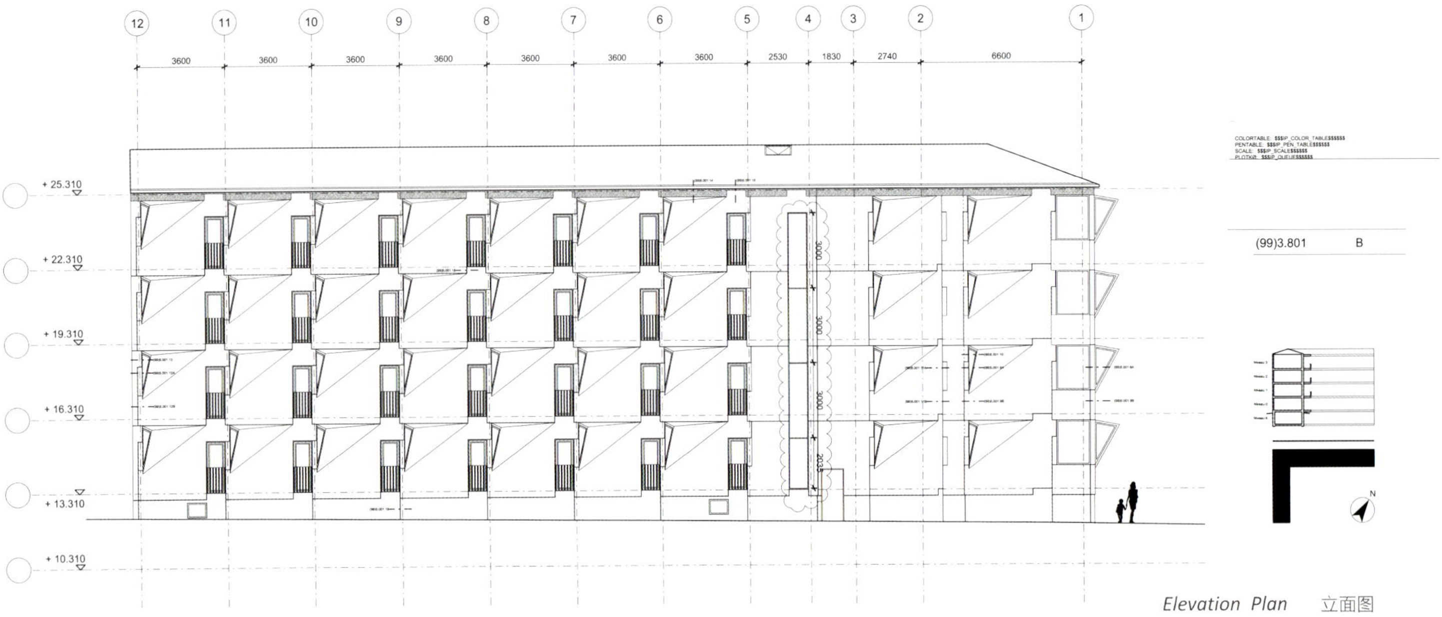

Elevation Plan 立面图

FEATURE 特点分析

Façade

Different views forms contrast on the façade by the bay windows. Softness of hard materials and changing light and shadow achieves great with doing little.

立面

凸窗造成了不同视觉方位上的立面对比，硬质材料的软化及光影变幻起到了四两拨千斤的整体效果。

一直以来，夹心板材构筑的建筑都被冠以"闷箱子"的外号，而 KHR 建筑设计公司的设计师却在菲特烈堡的 Hedorfs 学生公寓项目中证明了水泥构造的建筑也能充满创意和灵感。

受菲特烈堡传统公寓的凸窗所启发，KHR 建筑设计公司设计的几何形的凸窗创造了一个富有表现力的动态立面。凸窗高出楼面而砌，使公寓看起来更加宽敞，并富于动感。

在宽混凝土墙前的凸窗上方，设计师还设计了一个小三角窗，上面的倒影使建筑显得小巧精致。根据你观察位置的不同，建筑的外观也会有所不同，这些窗户产生了万花筒般的效果。从一面看立面开满了窗，而从另一面看，窗户又似乎全部被关上，这种有机变化使整个立面独具特色。由于凸窗的作用，整栋房子不需要灯光就产生了光影中的全天候色彩变化，立面的波动也使硬质材料外观显得柔软和自然。

Hedorfs 学生公寓毗邻 Flintholm 列车站，同时靠近哥本哈根商学院和布鲁克商学院。公寓有 4 层楼，一共 52 个单人间。每层有一个公共厨房，学生可以在那里碰面、烹饪和交流。

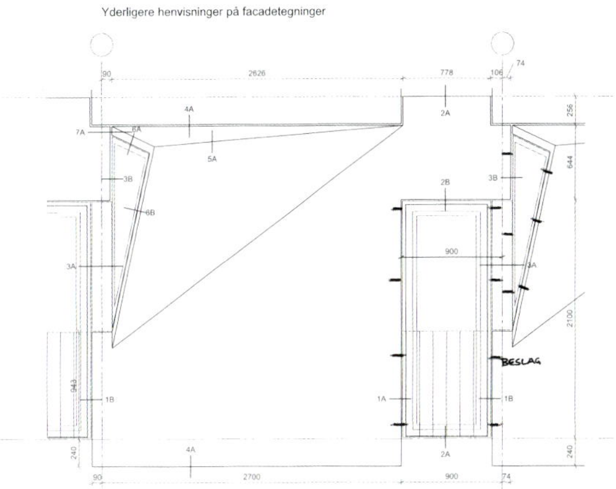

Detail Plan 节点图

Detalje 6A

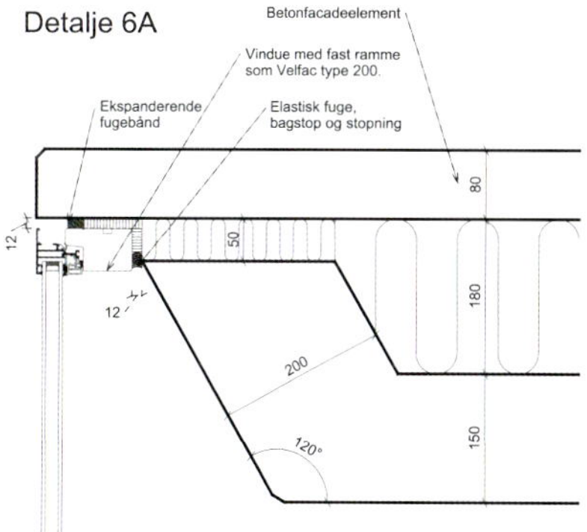

Detalje 6B

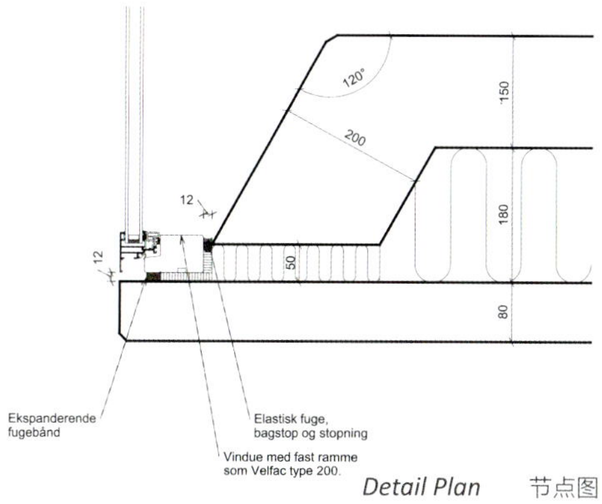

Detail Plan 节点图

Vandret snit ved afdækningsplade

Detalje 3A

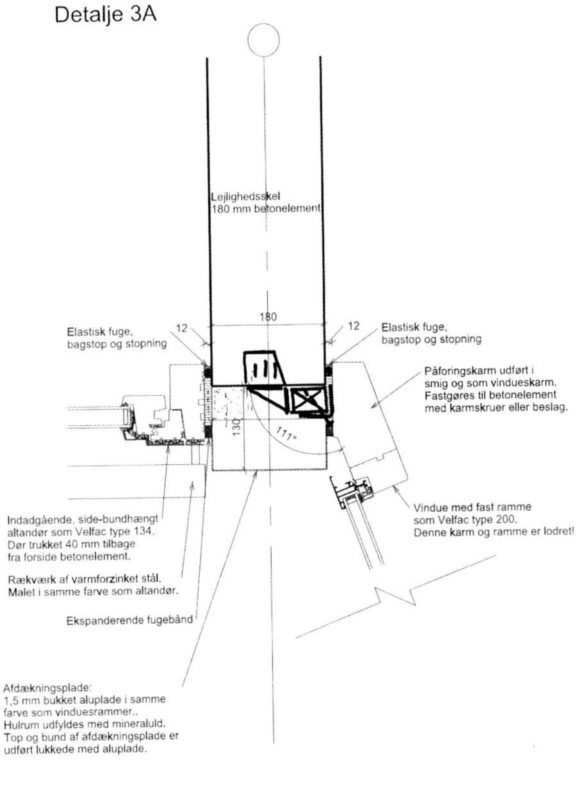

Vandret snit ved betonforplade

Detalje 3B

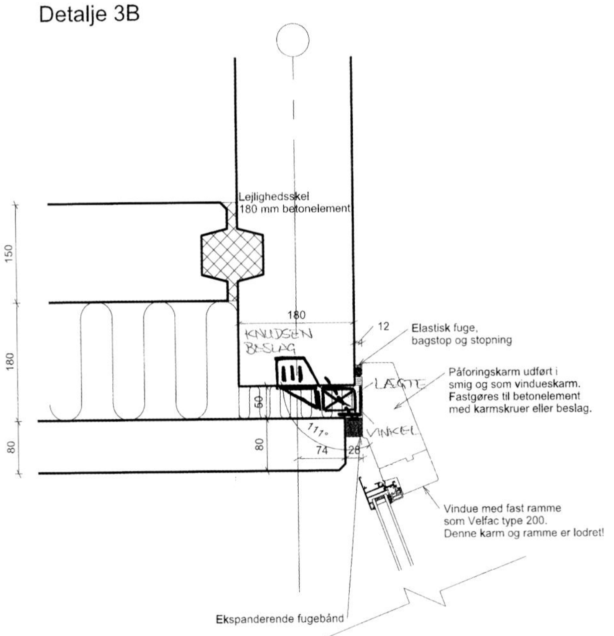

Detail Plan 节点图

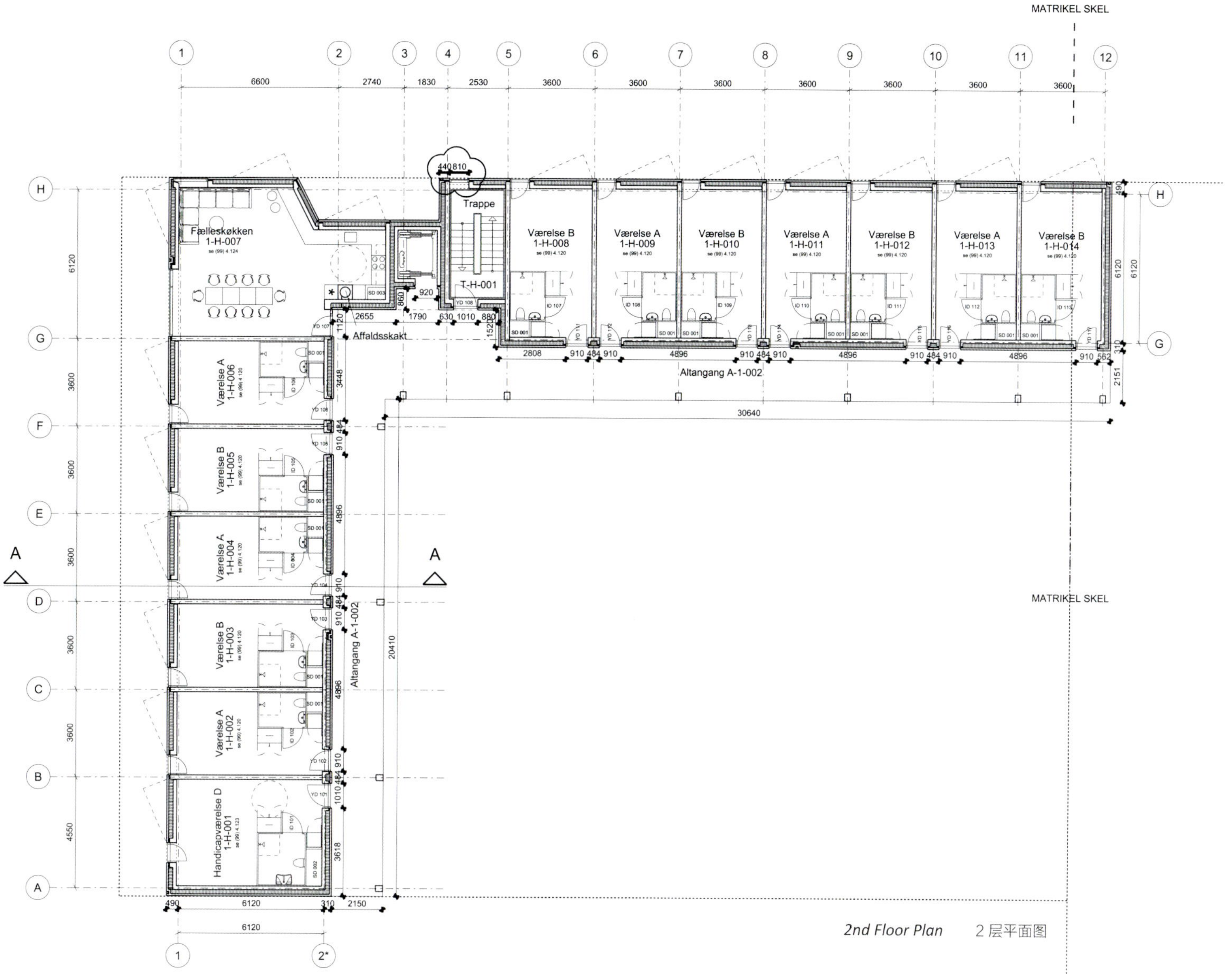

2nd Floor Plan　2 层平面图

One & Ortakoy, Istanbul, Turkey

Location: Ortakoy, Istanbul, Turkey
Architect: GAD & Gökhan Avcıolu
Site Area: 12,000 m²
Gross Floor Area: 56,100 m²
Number of Floors: 6
Photographer: Tahsin Inanici

土耳其伊斯坦布尔市 One & Ortakoy 住宅综合楼

地点：土耳其伊斯坦布尔奥尔塔科伊
设计公司：GAD & Gökhan Avcıolu 建筑事务所
占地面积：12 000 m²
建筑面积：56 100 m²
层数：6
摄影师：塔西恩·伊那尼斯

Surmount 突破点

SHAPE 造型

VOLUME 体量

FACADE 立面

ROOF TERRACE 屋顶花园

ORT is a mixed use complex currently under construction in the Istanbul neighborhood of Ortakoy. The project consists of two buildings; one building is residential and the second is commercial and retail. Nestled next to a hillside, the project will be an iconic step forward in the modernization of the area.

The project's form, façade and overall organization developed through a series of experimental strategies influenced by contextual elements. GAD's design solution optimizes the use of the various programs distributed throughout the site. The roofs on both buildings are intended to be landscaped recreational playgrounds. These green roofs lessen the scale of the buildings, merging them with the existing topography. As well, the natural stone façade that wrap both buildings provides a soft skin blending with the natural hillside setting.

该项目位于伊斯坦布尔的奥尔塔科伊，坐落在一个山坡边上，是一栋多功能综合大楼，目前还在建设中。项目包括两栋大楼，一栋是住宅楼，另一栋是商务和零售楼，是该地区现代化进程中具有标志性意义的一个项目。

该项目的结构形式，立面和整体组织都是通过一系列受环境因素影响的实验性策略展开的。相对于分布在这一区域内的其他各个项目，GAD & Gökhan Avcıolu 建筑事务所的设计方案可谓是其中的佼佼者。两栋建筑的屋顶被精心打造成景观优美的休闲游乐场，这种绿色屋顶花园减少了该项目的建筑规模，使它们与现有的地形完全融合。此外，建筑外部包裹着的天然岩石立面像柔软皮肤一样使整栋建筑与四周山坡的自然环境交融在一起，浑然天成。

FEATURE 特点分析

Volume

The solution is arranging the so given fixed-volume according to the needs of the containing shell. When there is a space using more volume the other spaces reacts, shrinking to an optimal value. Multiple parameters such as massing and arrangement, enclosure of the environment, harmony with topography and landscape as well as the prospect for each unit affect the design.

体量

建筑的整体设计根据规定的最大固定体量进行建设。当一个空间占用更多的体量，其他空间会产生相应的变化，把自身缩小到一个最优值。许多建筑参数例如构成与安排、环境的围合，与地形及景观的和谐，还有每个单元的视野都会影响整个设计。

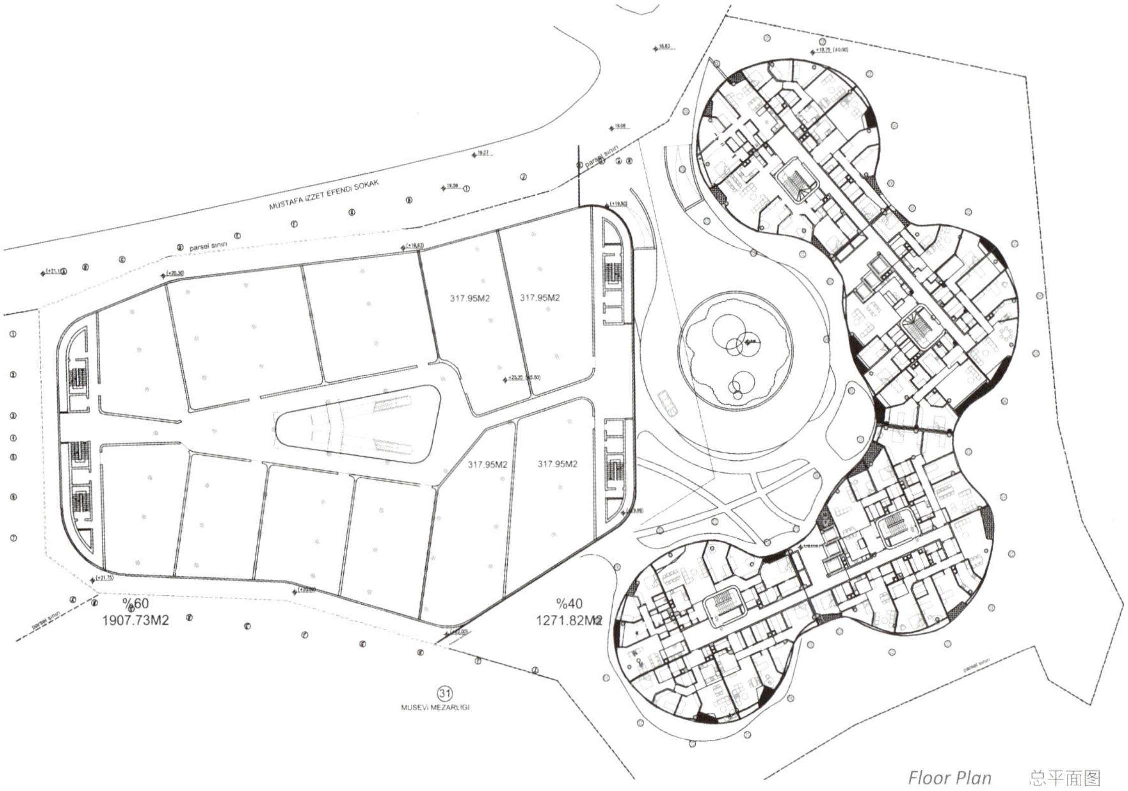

Floor Plan 总平面图

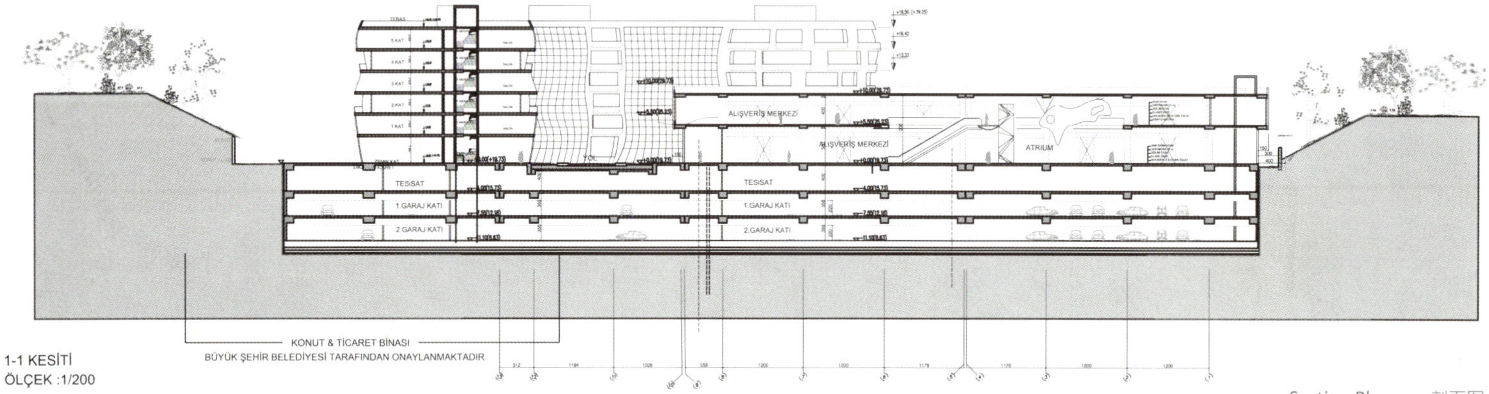

Section Plan 剖面图

D2 124.65 m^2 House Plan　D2 124.65 m^2 户型图

D2 164.40 m^2 House Plan　D2 164.40 m^2 户型图

D1 131.24 m^2 House Plan　D1 131.24 m^2 户型图

D1 147.16 m^2 House Plan　D1 147.16 m^2 户型图

D3 106.22 m^2 House Plan　D3 106.22 m^2 户型图

D3 127.72 m^2 House Plan　D3 127.72 m^2 户型图

4 Towers Osdorp, Amsterdam, The Netherlands

Location: Amsterdam, The Netherlands
Architect: Wiel Arets Architects
Gross Floor Area: 58,988 m²
Number of Floors: 12
Photographer: Jan Bitter

荷兰阿姆斯特丹市奥斯德普 4 Towers

地点：荷兰阿姆斯特丹
设计公司：Wiel Arets 建筑事务所
建筑面积：58 988 m²
层数：12
摄影师：简・彼特

Surmount 突破点

WINDOW 窗

LAYOUT 布局

FACADE 立面

ECONOMIC 经济

In Osdorp, southwest of Amsterdam, major changes took place to transform a vacant zone into a residential district connected to the existing housing area alongside it. One part of the residential district is a public garden, Jan van Zutphenplantsoen, where the 4 towers stand along a newly created water-park. All 4 towers are sited according to the given urban footprint and take up the maximum envelope, giving a total of 390 apartments evenly distributed over 11 levels.

Bringing a large number of units to the marketplace at once, the project questions the stacked repetition customary in such developments. Each unit has a patio that can open over the whole length creating a bright, covered outdoor space. Because of its position within the apartment, the living space, which already seems wider than it is, can be enlarged physically by folding away the dividing glass. In some instances a double-height space breaking though the stacked levels extends the living space vertically.

Three overlaid patterns create an elevation that conveys the internal organization to the outside world. First, an aluminum cladding indicates the separation between the apartments. Second, the mirroring concept of the plan results in a playful façade pattern ranging between alteration and iteration, between standardizing and individualizing. Third, the twofold elements bring about a subtle discontinuation within the unitized surface.

The towers share an underground parking facility lit and ventilated by 4 enormous voids. By vegetating the roofs of the parking facilities, the complex is made part of the adjoining public garden whose pleasant atmosphere is drawn into the glazed entrance lobbies. As the parking strip and entrance level are less wide than the main mass of the buildings, the cantilevered towers overhang adjacent conditions, both street and water.

阿姆斯特丹西南面的奥斯德普区已经发生了巨大变化，原有的大片空地被改造成与旁边的旧住宅区相连的新住宅区。该住宅区中有一个叫佐特芬普兰特松的公共花园，而 4 栋高楼就矗立在这里一个新建的水上公园里，临水而立。4 栋高楼按照规划选址，最大限度地利用其面积，建造了 390 间公寓，平均分布在 11 层楼上。

长久以来，大量同类项目不断涌现，却缺乏新意，而此次该项目的大量引进，必然给市场带来一定的冲击。每个单位都有一个全长的露台，形成一个明亮的、带屋顶的户外空间。拉开玻璃门后，原本已经看起来比实际要大的生活区，在视觉上会显得更大。而有些复式的跨层公寓则在垂直方向上扩展了其原有的生活空间。

外墙的覆盖样式有三种，每一种都能反映整个建筑的内部结构。首先，铝制面板的使用清晰界定了两个公寓之间的分隔。其次，镜像概念的设计使建筑外观充满趣味，既多变又单一，既符合标准，又独具个性。再次，双重元素的综合运用为原本单元化的外表带来变化，体现微妙的不连续性。

这 4 栋高楼共享一个地下停车场，停车场有 4 个大通风口采光和透气。停车场上面种植了大量的花草树木，使其自然地成为公共花园的一部分，为进出口大堂增添了靓丽的风景线。由于停车带和入口比建筑主体的宽度要窄，因此，这些支架式的建筑不得不悬于毗邻的街道和水上公园之上。

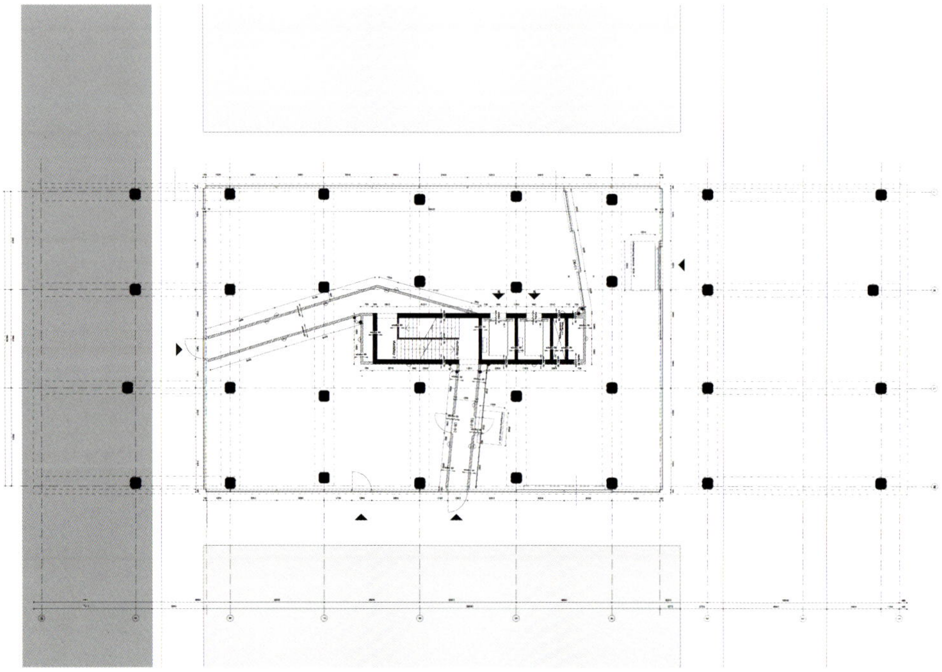

1st Floor Plan　　1 层平面图

FEATURE 特点分析

Layout

The plan allows for a high degree of interchangeability of layouts so as to offer a diversity of apartment types. A central core and a number of strategically positioned service shafts enable the different types to mirror and shift not only with regard to each other but also in terms of the spatial configuration within one apartment.

布局

此项目的布局形式新颖多变，是一间多元化的现代公寓。中央核心和规则放置的服务轴使得不同类型的布局以及不同的公寓内部空间配置能实现互换和切换。

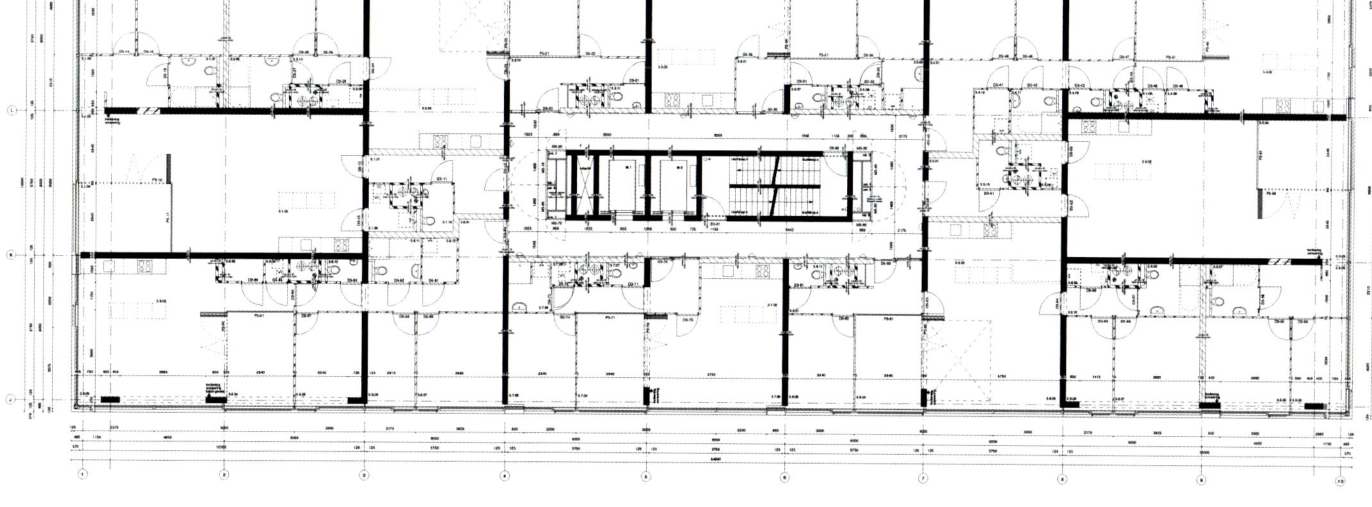

Floor Plan 楼层平面图

Residential Building C, Nova Gorica, Slovenia

Location: Majske poljane, Nova Gorica, Slovenia
Architect: Ravnikar Potokar ArKitekturni
Gross Floor Area: 8,200 m^2
Number of Floors: 11
Photographer: Peter Krapež

斯洛文尼亚新戈里察市公寓楼 C

地点：斯洛文尼亚新戈里察市 Majske poljane 住宅区
设计公司：拉弗尼卡尔 · 珀特卡尔建筑事务所
建筑面积：8 200 m^2
楼层：11
摄影师：彼得 · 卡拉佩斯

Surmount 突破点

BALCONY 阳台

SHAPE 造型

MATERIAL 材料

FACADE 立面

In line with the Decree on the Location Plan for the site along the railway station in Nova Gorica, Phase 2.a of the construction encompasses the construction of three 11-storey tower blocks, beneath which the construction of an underground garage is foreseen on the entire site of this phase. Tower Block C is located in the future building complex of the residential area Majske Poljane along the railway station in Nova Gorica. Within this phase of construction, 3 tower blocks will be erected along Prvomajska Street.

The architectural design of Tower Block C is derived from the competition design, as outlined in the study submitted to the competition in 2002 (which was subsequently awarded first prize in the competition) and in the adopted Decree on the Location Plan. The common vertical building height with a ground floor and 10 storeys also includes a gallery on the ground floor. The ground floor and gallery are intended for public use, for various service providers and office activities, and for grocery stores and other shops, while the upper floors are intended as flats (the tenth floor features two-storey duplex flats). The building has two basements intended for technical or maintenance facilities, storage rooms belonging to the flats and parking spaces for cars.

RAZLOGOM
Umek & Carl Cox
v parku Tivoli

根据政府的地区规划，新戈里察市火车站旁的第 2a 期建设工程包括 3 栋 11 层高的建筑和一个地下停车场。住宅楼 C 位于新戈里察市火车站旁的 Majske Poljan 住宅区未来的综合楼群中。这期建设完成后，3 栋塔式大楼将会伫立在 Prvomajska 大街上。

住宅楼 C 的设计来自 2002 年设计大赛中提交的参赛作品（随后在竞争中荣获一等奖），后来被政府采纳到地区规划中。建筑的垂直高度包括底层和其上的 10 层，这 10 层又有一个走廊作为底层。底层和走廊计划为公用区域，为各种服务供应商、商务活动、零售店和其他商店等提供场所，上面的楼层则是公寓（第 10 层楼为复式套房）。建筑的两个地下室分别用来摆放停车场的技术维护设施，用作公寓的储物室和停车场。

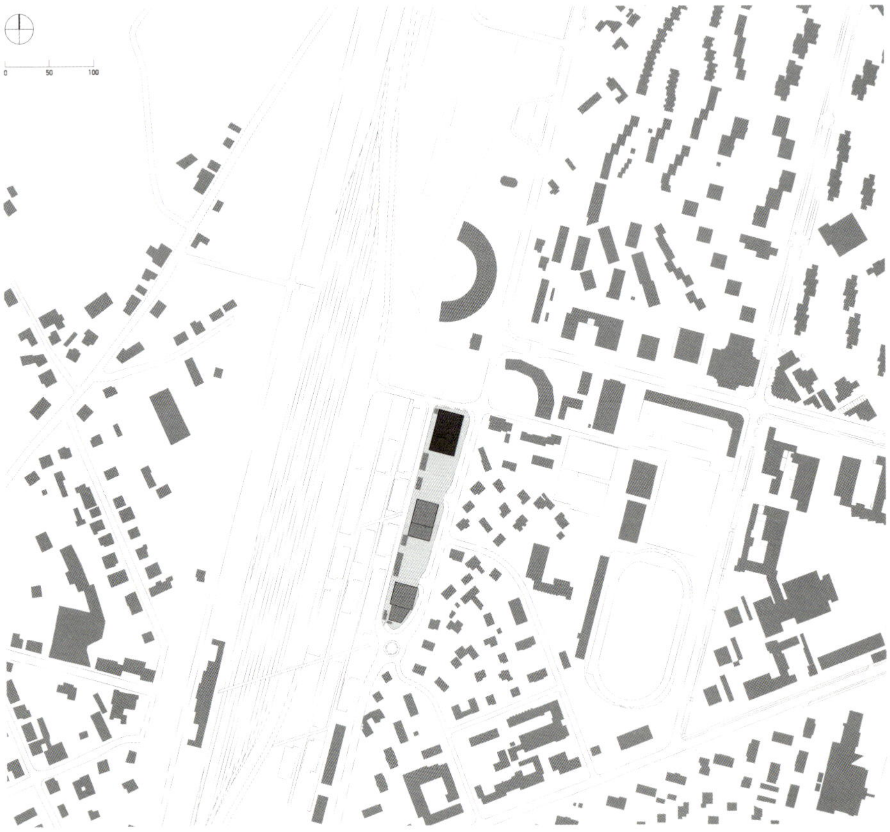

Site Plan 区位图

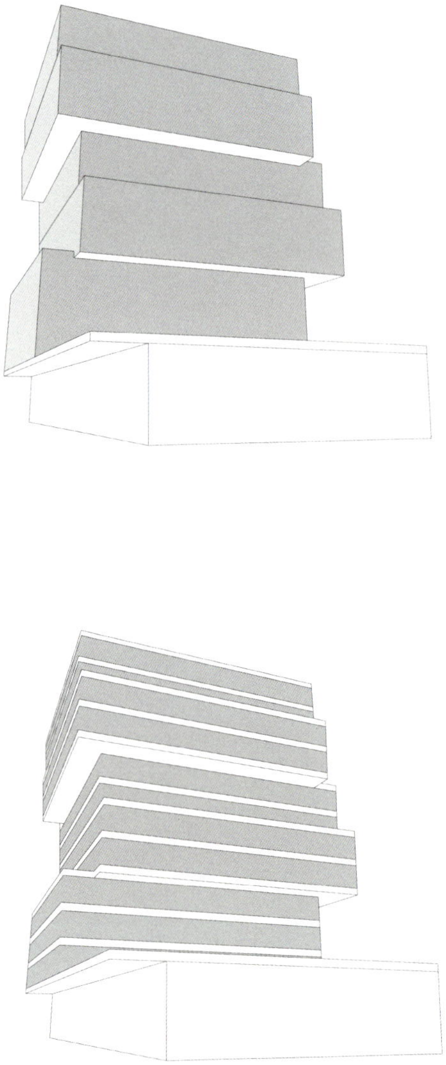

Axonometric Drawings 轴测图

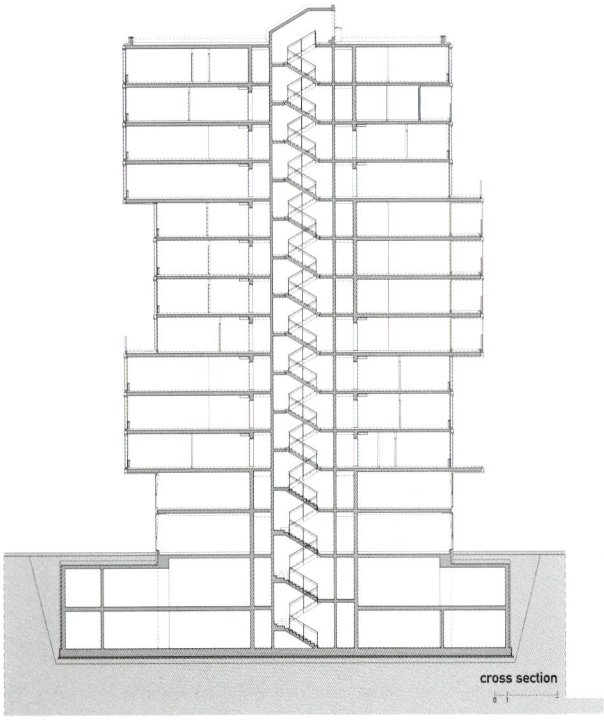

Cross Section Plan 横截面图

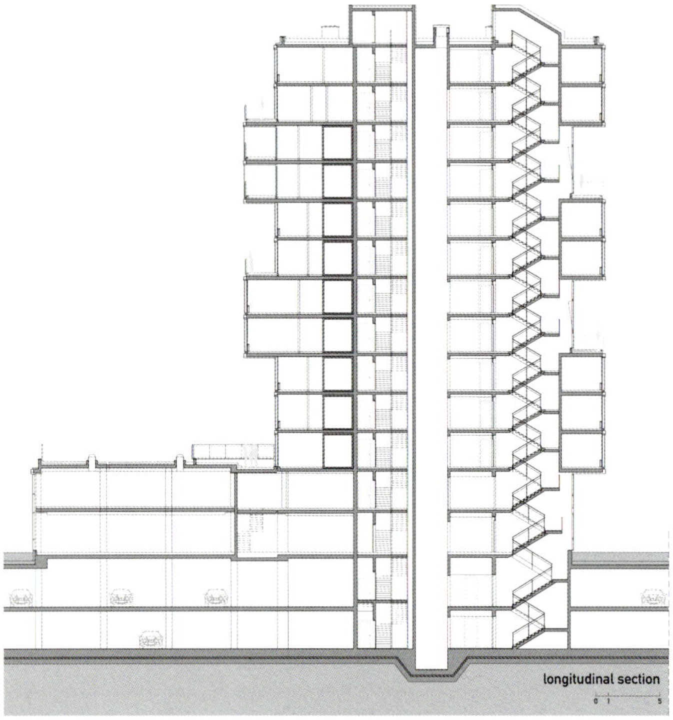

Longitudinal Section Plan 纵向剖面图

FEATURE 特点分析

Shape

The basic idea lies in the layering of two storeys by displacing their floor plan – the elementary square of the floor plan is moved around the vertical core thus creating overhangs on which terraces are placed. The vertical line of the residential tower is cut up and transformed into horizontal layering, providing the tower with the necessary dynamics.

造型

基本的设计思路在于两层楼之间楼层的错开——方形楼面围绕着中轴移动，产生了突出，阳台就建立在这些突出之上。建筑的中轴线被切断，转换成横向分层，使建筑看起来更具动感。

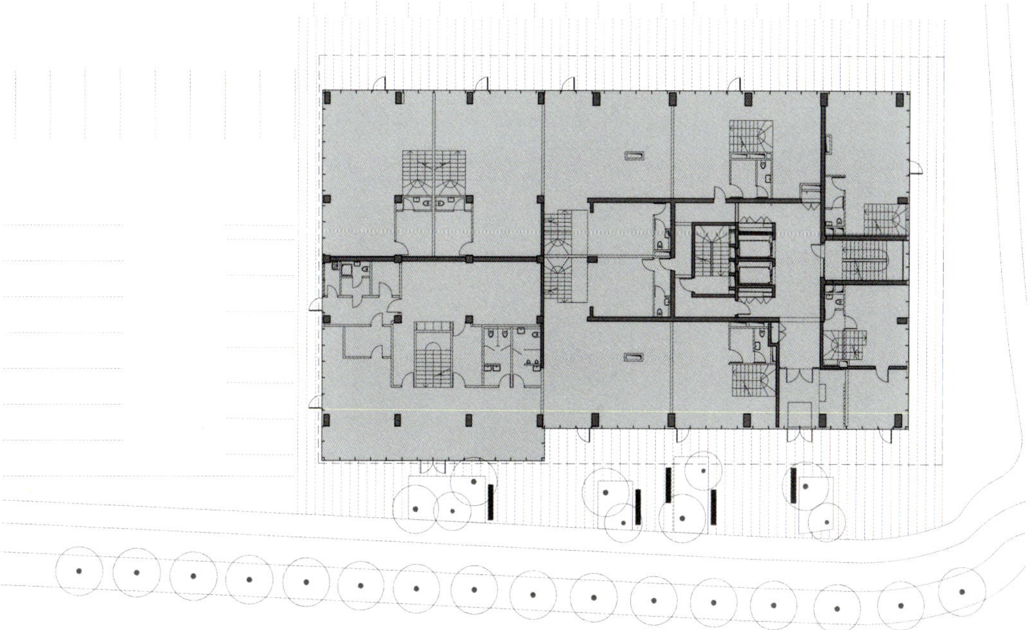

1st Floor Plan 1层平面图

Filadelfia Corporate Suites, Mexico City, Mexico

Location: Mexico City
Architect: Bunker Arquitectura
Gross Floor Area: 2,300 m²
Number of Floors: 5
Photographer: Héctor Armando Herrera & Fabiola Menchelli

墨西哥墨西哥城菲拉德尔菲亚商务酒店

地点：墨西哥城
设计公司：邦克建筑事务所
建筑面积：2 300 m²
层数：5
摄影师：埃克多·阿曼多·埃雷拉、法布瓦拉·曼士力

Surmount 突破点

WINDOW 窗

MATERIAL 材料

FACADE 立面

COLOR 色彩

What happens when the concept of a project is reduced to a façade in the process of its development? Are purely aesthetic aims valid in architecture? Is façadism something worthwhile? Are the designers to become architectural dermatologists?

A client contacted the designers to design a hotel for corporate suites in Napoles, a residential neighborhood in Mexico City that has been rapidly converting to office architectse over the past few years. The plot was located across the street from the convention center of the World Trade Center, by far the barchitectsiest office building in the city. The suites were intended to accommodate barchitectsinesspeople visiting the WTC and the year round expositions.

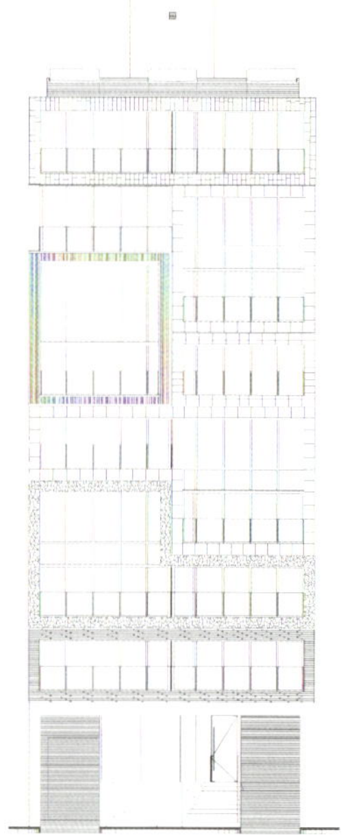

North Elevation 北立面

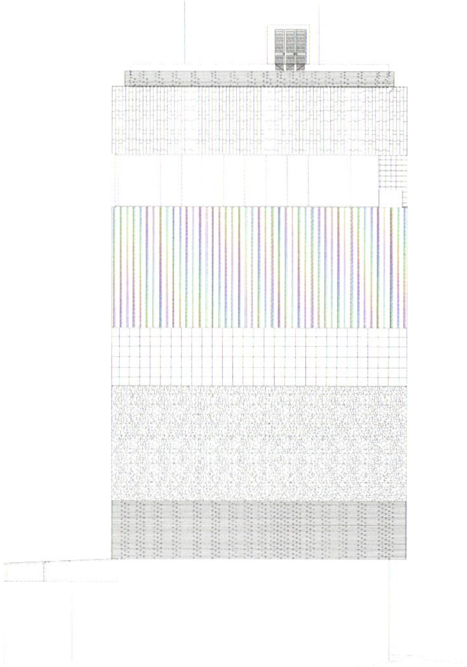

East Elevation 东立面

One of the biggest complaints of caffeine-charged traveling barchitectsinesspeople is that no matter how hard they try to avoid it, they must wake up at 5:30 a.m., take a cab to the airport, a Starbucks coffee while waiting at the gate, a tasteless omelet in the plane, a cab to the same forgettable hotel, meeting at 9 a.m., barchitectsiness lunch, afternoon meetings, drinks, back to the hotel, sleep, wake up at 5:30 a.m., catch the flight back... and so on. Their lives seem filled with sameness.

This motivated the designers to create a hotel where every recurring visit would be a completely different experience in an attempt. The different shapes were then assembled like a giant Tetris to form a vertical tower. The tower culminates with the last shape left in a cantilever to generate sarchitectspense... The assembled shapes were then raised to provide an entrance to the building.

When the construction of the building had begun, the client suddenly changed his mind and decided that from an operational point of view

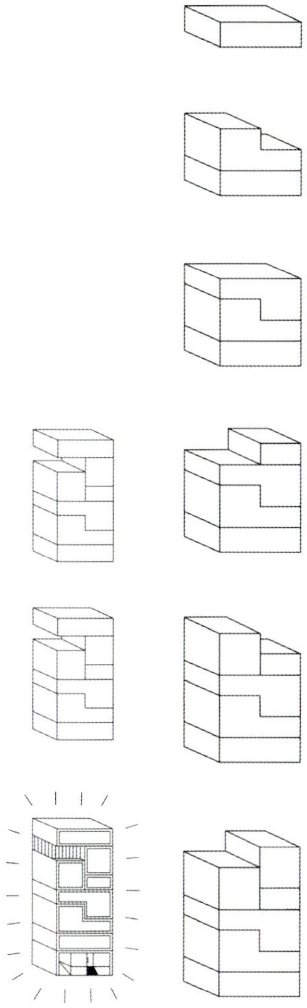

Volume Massing Diagram 体量构成示意图

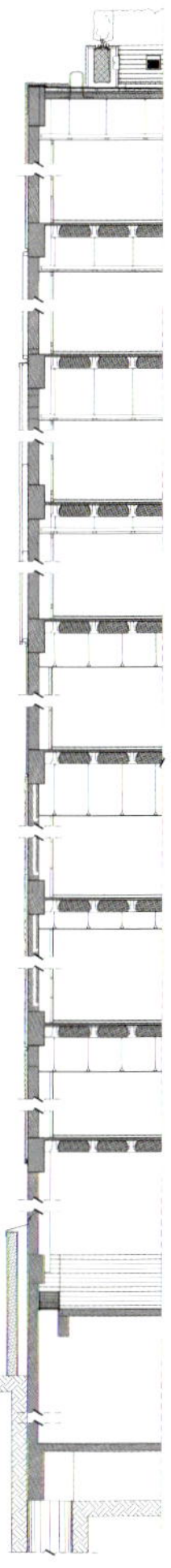

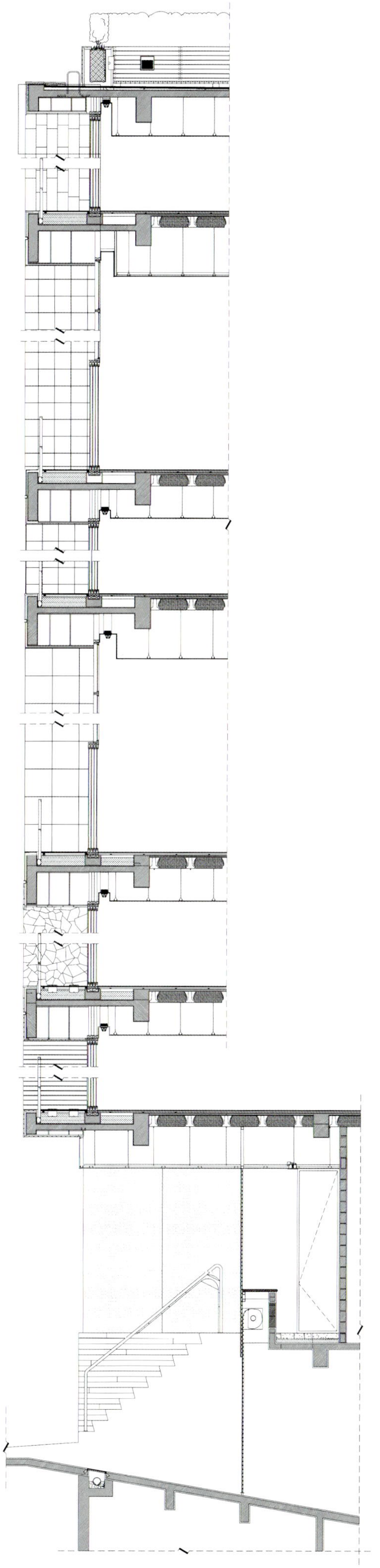

Details 节点图

it was too complicated to have 15 different suites. He wanted only two types of suites, the single and the double height. We were shocked.

He encouraged architects by letting the designers know his decision was bitter-sweet. The bitter part was losing the original concept and the sweet part was that the designers could maintain the façade becaarchitectse he really liked it. The designers had to find a way to adapt the floor plans and reorganize the interiors without sacrificing the façade. In the end, the client got his hotel and the designers kept our façade.

当一个项目的设计理念在建造过程中被削弱成只集中于立面的呈现会怎么样呢？单纯地追求建筑审美真的有效吗？立面主义值得追求吗？设计师岂非要成为建筑皮肤科医师了？

一个客户邀请设计师设计一家位于 Napoles 的酒店及其商务套房，Napoles 是墨西哥城的住宅区，在过去的几年间，这里迅速发展并转变成商业区。该地块与目前世界上最繁忙的办公楼——世界贸易中心会展中心隔街相望。而套房就为造访世界贸易中心和参加展览会的商界人士提供住宿。

商界人士因为经常旅行，只能靠咖啡因提神，因此他们最大的抱怨之一就是无论怎么努力去避免，他们的生活始终千篇一律，早上 5:30 起床，乘坐出租车到机场，候机时喝一杯星巴克咖啡，在飞机上吃无味的煎蛋卷，9 点开会，中午吃商务餐，下午开会，喝下午茶，回到宾馆，睡觉，早上 5:30 起床，赶回程航班……

这就是设计师设计这个酒店的动机：使每个经常来访的人每次都有一个完全不同的体验。不同形状的房间有序组合起来，形成一个像俄罗斯方块般的巨塔。酒店最高层的方块被悬空出来以形成悬念……最后，组合起来的方块被整体抬高，为建筑提供出入口。

建筑施工已开始，客户却突然改变主意，觉得从可行性方面看，15 个不同的套间太过于复杂。他只想要两种类型的套间——单层房和双层房就够了。这让设计师很为难。

他安慰说这个决定是两面的，虽然白费了最初的设计，但是他很喜欢这个立面，设计师可以保留它。因此，设计师设法修改平面图，重组室内布局，但又不破坏最初设计的立面。最后，客户得到了他的酒店，设计师也保持了设计的立面。

FEATURE 特点分析

Material

People's daily lives will always turn into a boring and tedious routine. To break their drudging routine, every suite would have a different floor plan and spatial arrangement and each shape was conceived of a different material: wood, metal, volcanic stone, marble, ceramic, limestone, glass...

材料

人们的生活一直都很枯燥、乏味，为了打破他们枯燥的日常规律，套房的特点需要进一步突出，每间套房都有不同的平面布局和空间安排，每个方块都用不同的材料打造，如木料、金属、火山岩、大理石、陶瓷、石灰石和玻璃等。

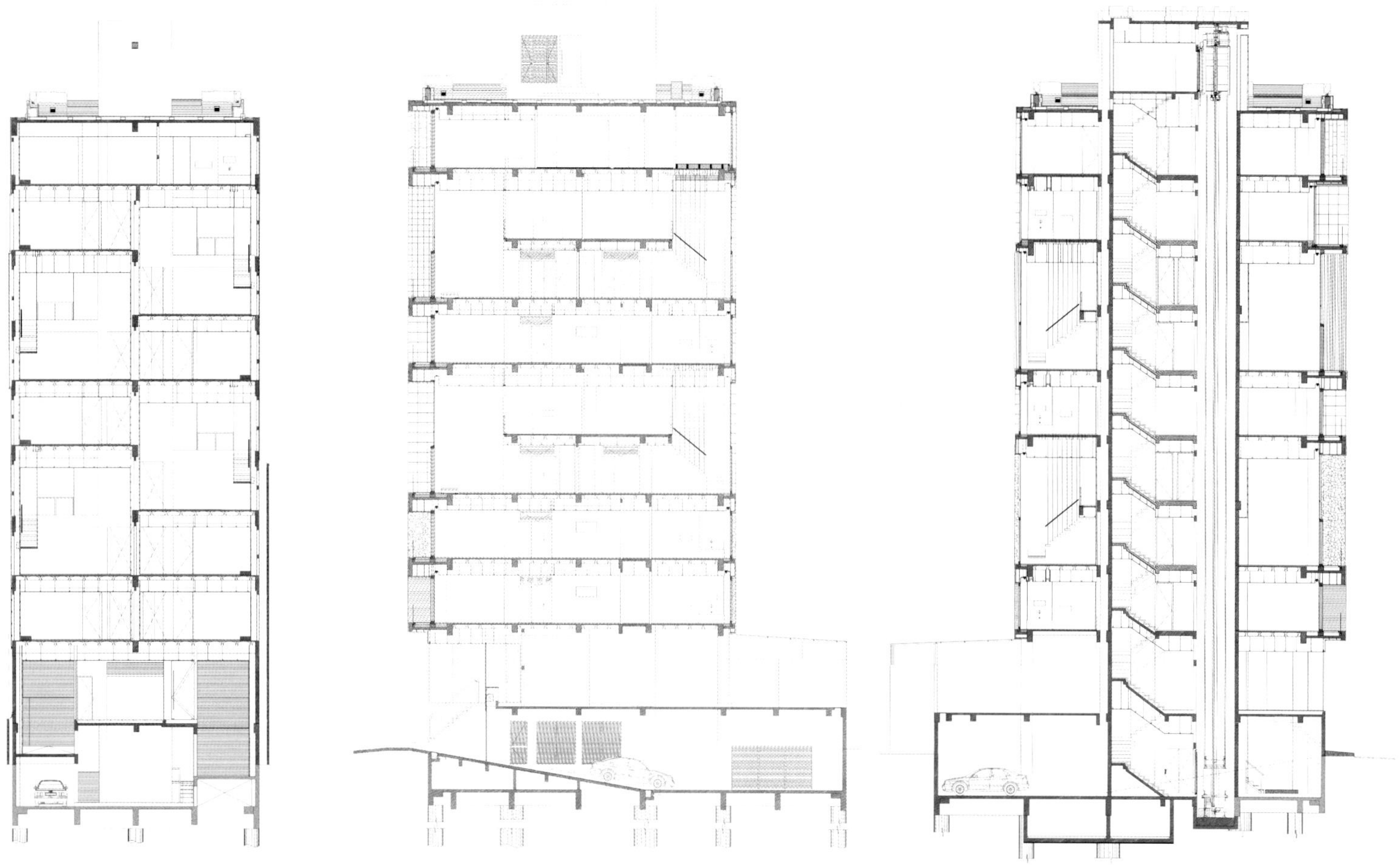

Section Plan　剖面图

7th and 8th Floor Plan　7 层和 8 层平面图

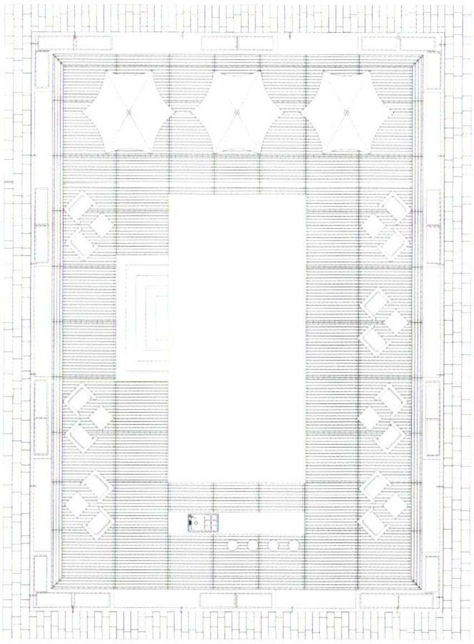

Roof Plan 屋顶平面图

Blok 1 Presikhaaf, Arnhem, The Netherlands

Location: Arnhem, The Netherlands
Architect: GROUP A Architects
Site Area: 3,000 m^2
Number of Floors: 10
Photographer: Scagliola/Brakkee, Rotterdam and GROUP A

荷兰阿纳姆市学生公寓楼1号楼

地点：荷兰阿纳姆
设计公司：GROUP A 建筑事务所
占地面积：3 000 m^2
层数：10
摄影师：鹿特丹 Scagliola/Brakkee、GROUP A 建筑事务所

Surmount 突破点

BALCONY 阳台

FACADE 立面

WINDOW 窗

COLOR 色彩

Studententower Blok 1 is part of the redevelopment, which proposes to partly demolish existing post-war housing estates in the Presikhaaf neighbourhood, making way for new, low density, housing developments. Blok 1 stands out in this development because of its high density character and its position in the existing ecological zone, serving as a landmark and visual beacon of renewal for the Presikhaaf area. With this in mind, GROUP A Architect designed a building that looks larger than it actually is but blends in with the natural surrounding.

Because of the tower's rectangular floor plan, its elevations vary in width. The narrow elevation gives the tower a tall and slender appearance, whilst the wider elevation functions as backdrop for the surrounding area, closing off the central axis in the Presikhaaf redevelopment scheme. To ensure that the building does not appear closed off and inaccessible, the façade is punctured by a number of large glass windows that reveal the interior of the building. This particular fenestration also ensures no two dorms are the same.

Blok 1 is the only building in Presikhaaf located within the ecological zone. GROUP A Architects therefore proposed a façade that interacts with its surroundings, and at the same time provides the design with a more layered meaning.

The ten story building is situated on an isolated site, surrounded by nature. The ground floor comprises support facilities for the students, technical areas and commercial units. Each of the 9 upper floors is laid out with 10 dorms. The top floor will be used as a communal roof terrace. Within the restricted budget, GROUP A Architects has been able to create a number of high quality solutions, like the large windows on the corners and the roof terrace with its nice views.

学生公寓楼 1 号楼是重建计划的一部分，该计划将部分拆除现存于普雷史哈夫社区的战后集合住宅群，为新型低密度的住宅开发项目创造条件。1 号楼因其高密度及其在生态区的地理位置而显得异常突出，它不仅会成为普利斯卡夫地区的地标，而且将会是该地区复苏的可视标志。考虑到这一点，GROUP A 建筑事务所设计了这栋看起来比实际体积大却又能和周围自然融合的建筑。

大楼长方形的平面使立面的宽度不同，狭长的立面使大楼看起来高而修长，同时宽阔的立面作为周边环境的背景，靠近普利斯卡夫地区重建计划的中轴线。为了使大楼看起来不显得封闭和难以接近，立面开了许多可以看到大楼内部的大型落地窗。这种特殊的开窗法使所有的宿舍都不尽相同。

因为 1 号楼是普利斯卡夫地区唯一一栋坐落于生态区内的建筑，GROUP A 建筑事务所设计了一个与周边环境互动的立面，并且赋予其更有层次感的含义。

这栋 10 层高的公寓楼孤单地矗立在一块被大自然所拥抱的地方。公寓楼首层由为学生提供服务的辅助设施、技术区和商业单位组成。楼上的 9 层楼中每一层都有 10 间公寓。顶层将用作公共屋顶天台。在有限的预算之内，GROUP A 建筑事务所依然创造出很多高水准的设计方案，比如说在角落处安设的大型玻璃窗和便于观赏美景的屋顶天台。

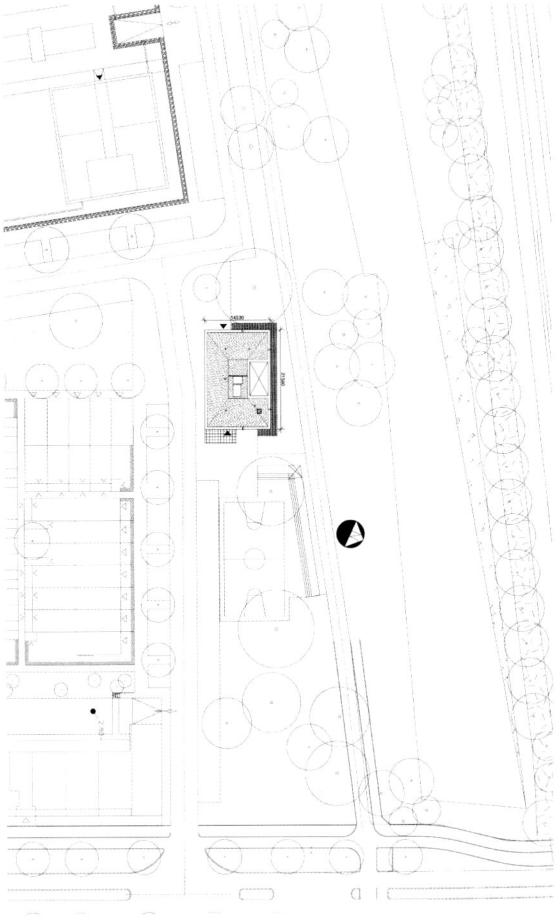

Site Plan 总平面图

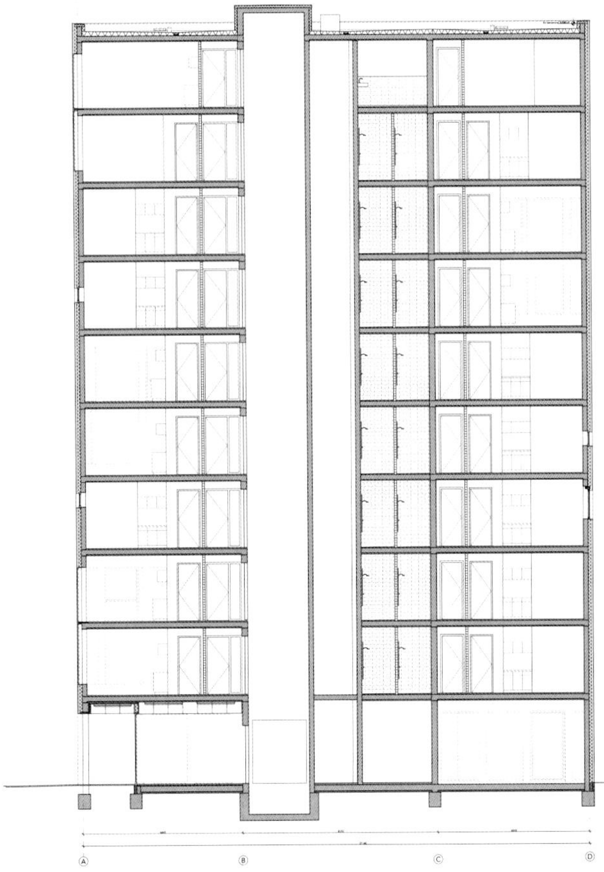

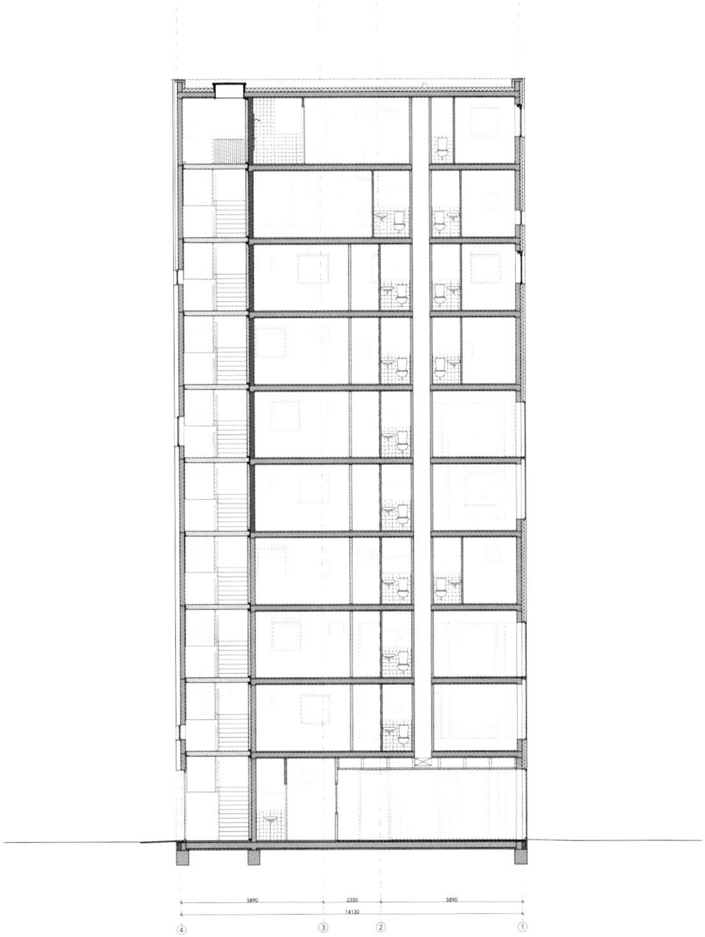

Section Plan 剖面图

Concept Plan 概念分析图

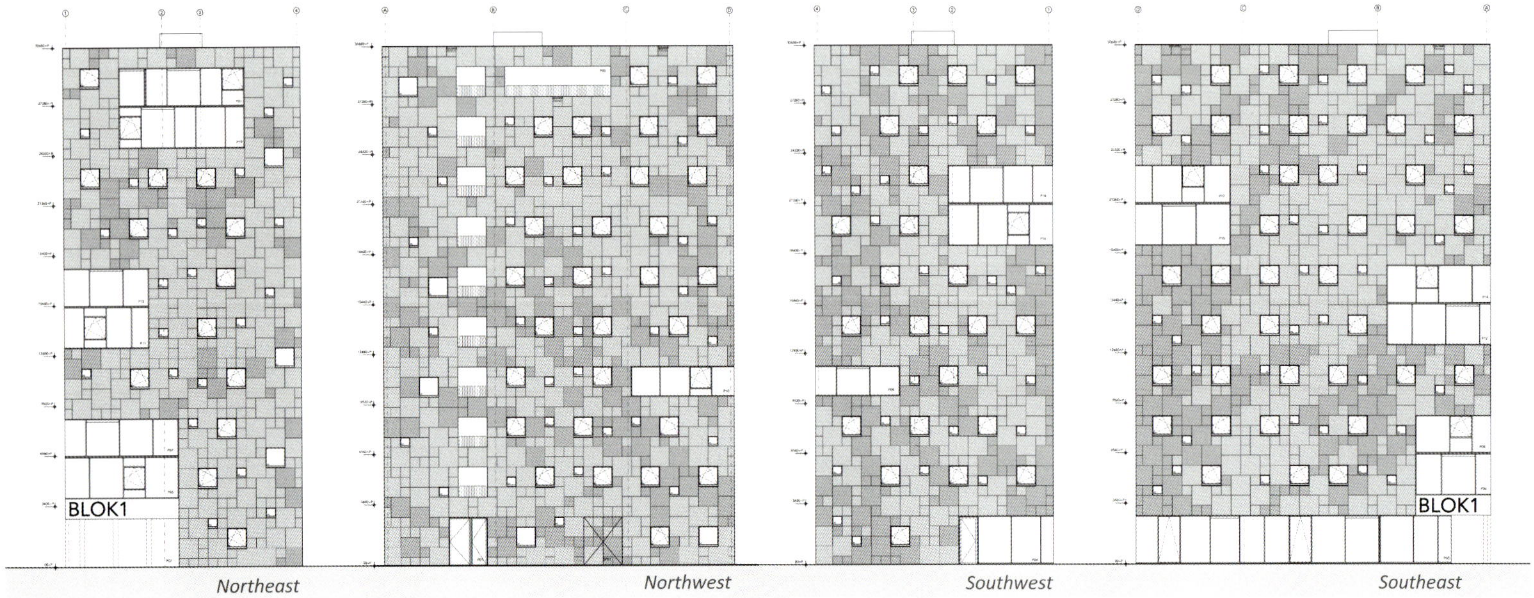

Northeast *Northwest* *Southwest* *Southeast*

Elevation Plan 立面图

FEATURE 特点分析

Façade

It's a building that looks larger than it actually is, because of the varying size of the window openings, which are "scattered" over the façade. The façade will be covered in paneling with a slightly differing finish, creating a subtle pattern that refers to the structure of a tree, without being literally recognizable. Not only the concept, but also the colors match the neighborhood perfectly.

立面

立面上不同规格的窗户，使建筑物看起来比实际体积大，立面由饰面上有细微不同的镶板覆盖，创造出一种微妙的树形结构格局，却不会轻易从表面被认出。除了设计概念，其颜色也与周边环境十分相配。

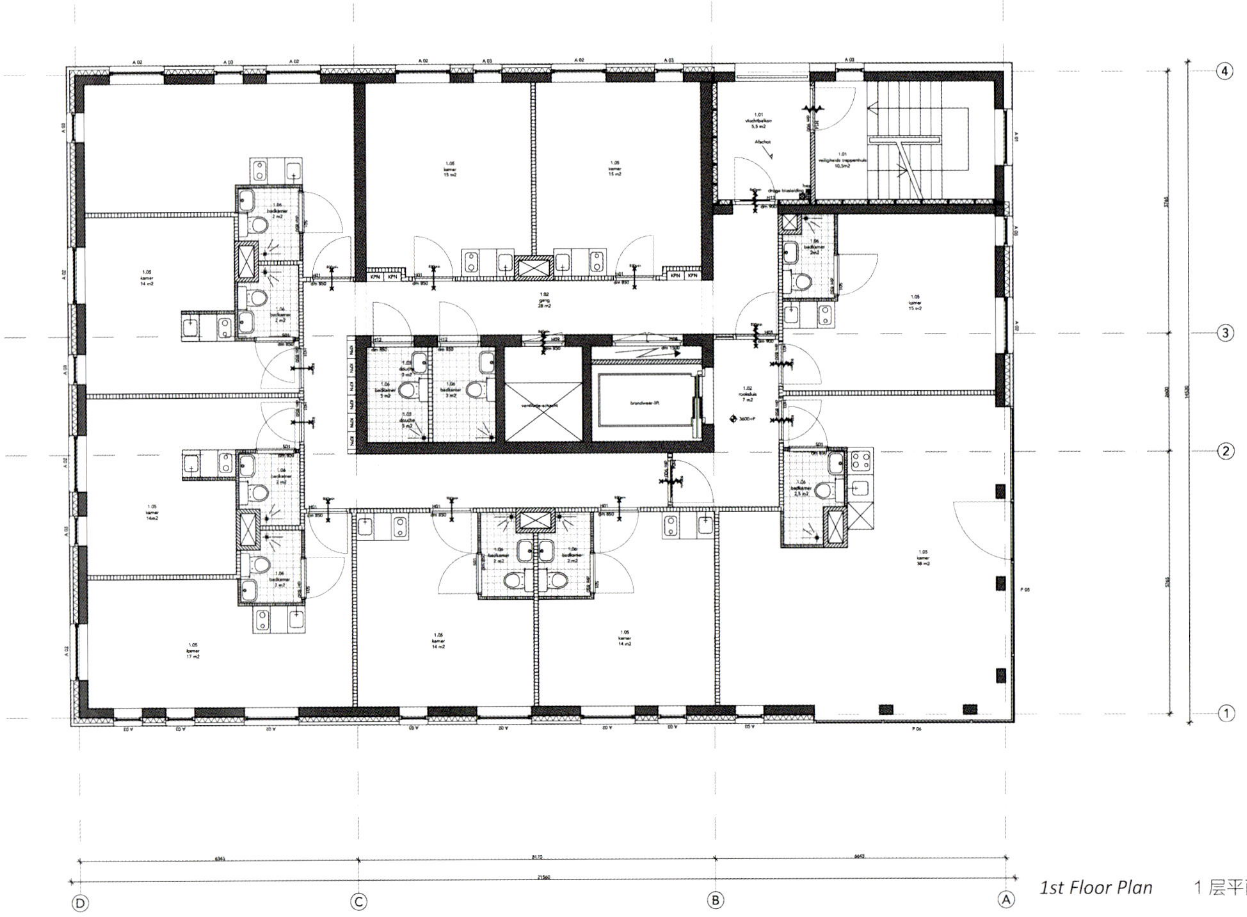

1st Floor Plan　1层平面图

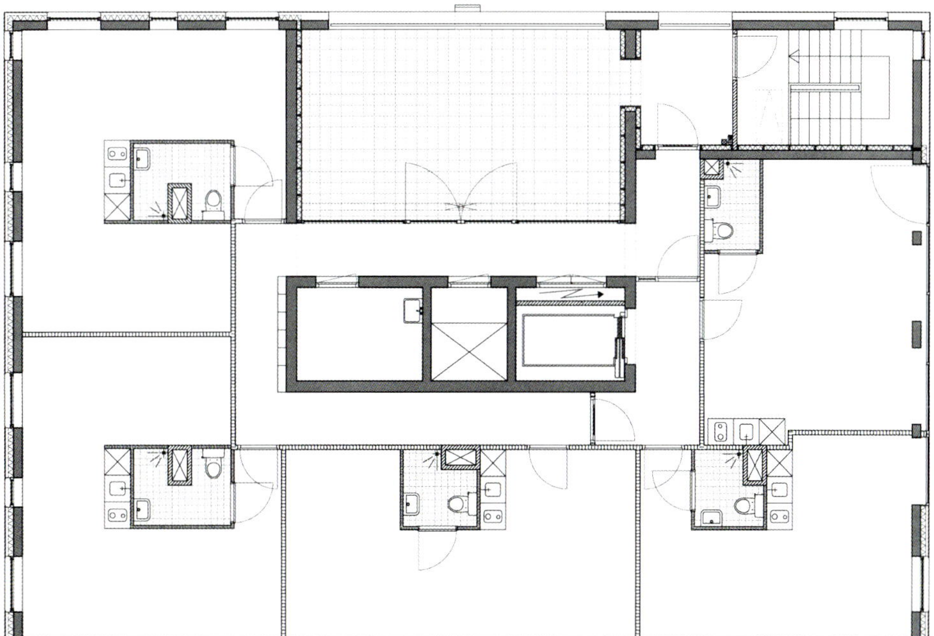

9th Floor Plan 9层平面图

The Silhouette, Aarhus, Denmark

Location: Løgten, Denmark
Architect: C. F. Møller Architects ,Christian Carlsen Arkitektfirma
Number of Floors: 11
Site Area: 3,000 m^2
Photographer: Julian Weyer
Prizes: Residential of the year WAN Award 2010, Aarhus Municipality Architecture Award 2010

丹麦奥尔胡斯港西罗町公寓楼

地点：丹麦勒格滕
设计公司：C. F. Møller 建筑事务所、克里斯蒂安·卡尔森建筑事务所
楼层：11
占地面积：3 000 m^2
摄影师：朱利安·韦尔
所获奖项：2010 WAN 年度住宅奖；2010 奥尔胡斯市建筑奖

Surmount 突破点

DAYLIGHTING 采光
SHAPE 造型
BALCONY 阳台
VIEW 视线

Many towns in Denmark have centrally-located industrial silos; most are no longer in use, but continue to visually dominate the local skyline. This is also the case with the town of Løgten north of Aarhus, where the former silo complex has been transformed into a "rural high-rise", with 21 high-quality residences composed as individual and unique "stacked villas".

This provides an alternative to standard apartments or detached suburban sprawl, in a mix of single-storey flats and maisonettes, in such a way that even the lower levels fully get to enjoy the views, and no two flats are the same.

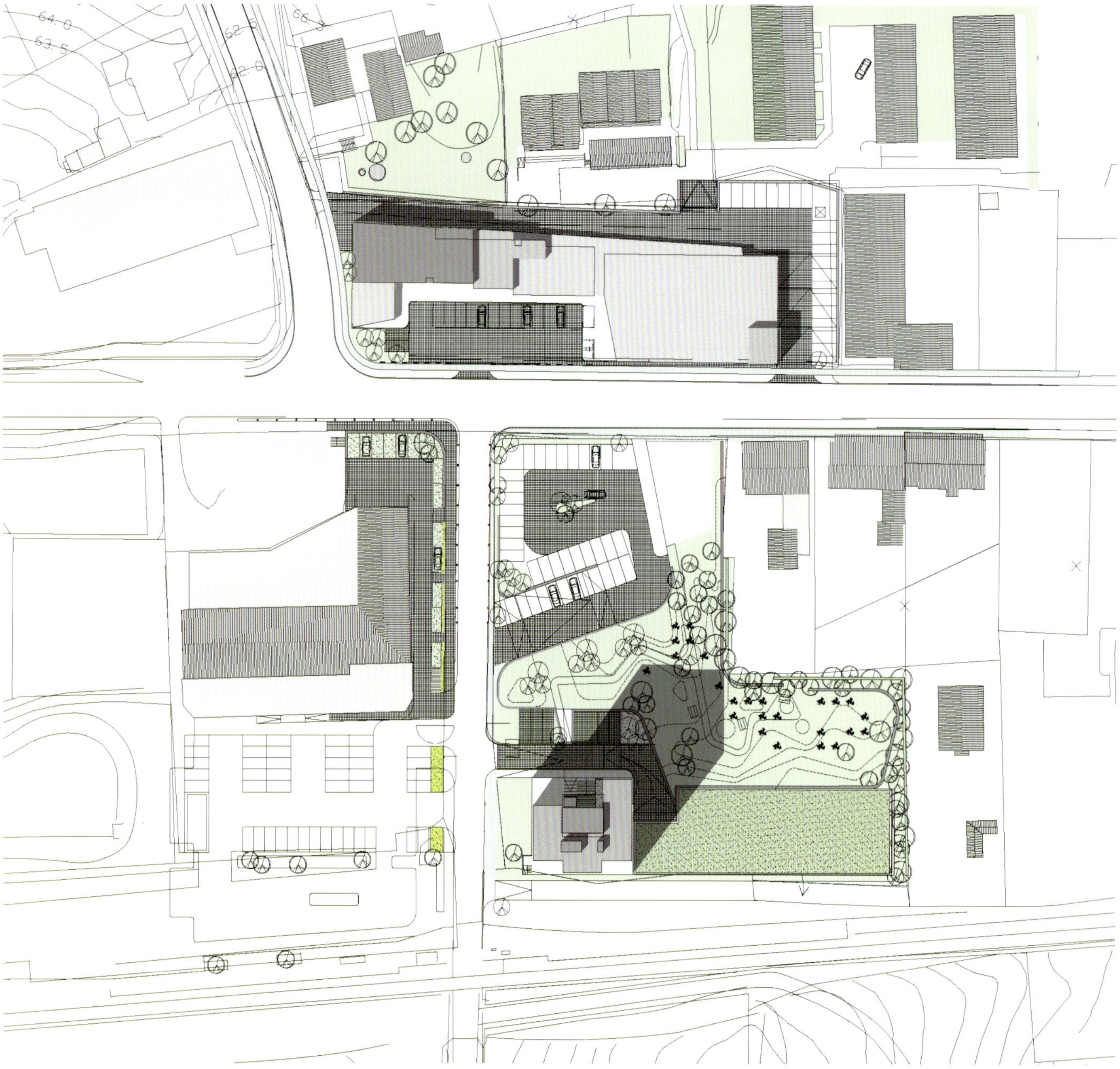

Site Plan 总平面图

The actual silo contains staircases and lifts, and provides the base of a common roof terrace. Around the tower, the apartments are in eye-catching forms which protrude out into the light and the landscape – a bit like Lego bricks.

Every apartment enjoys sunlight in the morning, mid-day and evening, whether placed to the north or south of the silo structure.

At the foot of the silo, a new "village centre" has been created, with a public space surrounded by a mixed-use complex with shops, a supermarket and terraced housing, and a green park containing small allotments for the residents.

The nature of the silo's "rural high-rise" remains unique – since it is a conversion, no other building in the area can be built to the same height, and it will remain a free-standing landmark. It is an example of how the transformation of redundant structures can both provide a new identity and introduce density to suburban outskirts.

The body of the silo has been deliberately left visible on the side of the building facing the new centre, to ensure a continued legibility of the history of the site, and to acknowledge that these types of structures have as much validity as rural historical markers as, for example, church bell towers or historic windmills.

过去丹麦有许多市镇都把工业仓库放在中心地区，现在它们大多数都已经被废弃了，但仍然是当地的天际线重要的组成部分。奥尔胡斯北边的 Løgten 镇也不例外。这里的仓库综合楼现在被改建成了乡村高层住宅楼，里面包含 21 套高品质住宅，以个性而独特的“复式别墅”形式组合在一起。

这种设计使住户可以在单层和复式住宅中自由地选择标准式公寓或郊区分离式住宅，即便是在低层的住户也可以欣赏到美景。此外，在这住宅楼中的公寓没有两间是完全相同的。

大楼里配有楼梯和电梯，还设置了基本的屋顶露台。围绕着大楼，里面的公寓以一种引人注目的结构外形，像一块巨型的乐高积木般闪耀在广阔的景观和交错的光线中。

此外，住宅楼筒仓式的建筑结构也使无论是在北边还是南边的公寓楼都享受到早晨、中午和傍晚时的阳光。

围绕着这座建筑，这里已经形成了一个“商业中心”，包括一个被内含商店的多功能综合楼所围绕着的公共区、一个超市和联排房，以及一个部分专为居民服务的绿色公园。

在这个区域里，没有其他建筑能建到同样的高度，相比从前，它是一个转变，且仍然是这里一个独立的地标。因此，对于这栋乡村高层住宅楼来说，它依然是独特的，它为其他建筑的改建提供了一个极好的示范，告诉人们该如何为这类残余建筑创造一种新身份，同时把人口引入郊区。

建筑的主体中正面对着新商业中心的一面被刻意地保留下来，以传承和发扬当地历史，让人们意识到这种建筑改造和教堂的钟楼或历史悠久的风车这些乡村历史标志的保留一样，都是对历史文化的有效保护。

FEATURE 特点分析

Shape

Around silo the tower, the apartment is built on the unusual steel structure with its protrusions and displacements. The unique shape provides all of the apartments with generous outdoor spaces, and views of Aarhus Bay and the city itself.

造型

围绕着大楼，里面的公寓皆以独特的钢结构为基础，以突出和位移的方式进行组合。这种独特的造型为公寓楼提供了宽阔的室外空间和良好的采光，使其中的居民们能纵览奥尔胡斯湾和整个城市的美好风光。

East Elevation 东立面图

West Elevation 西立面图

North Elevation 北立面图

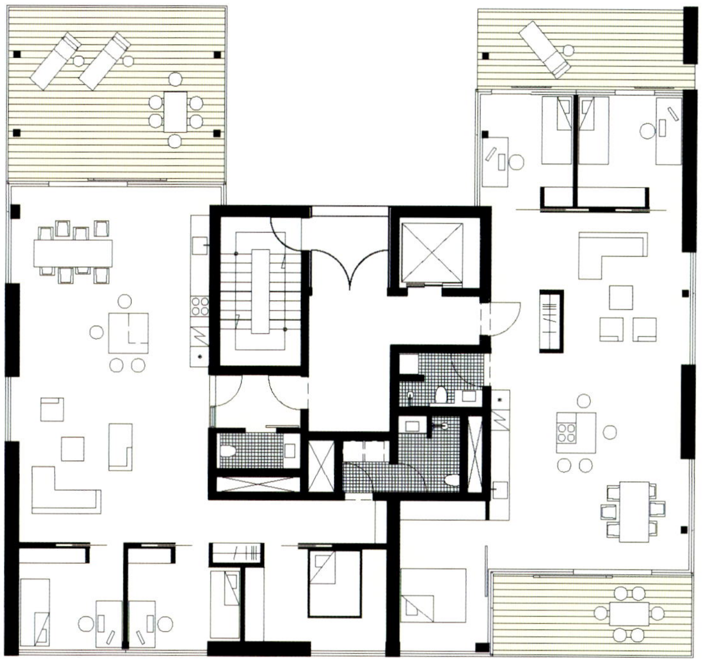

5th Floor plan　5层平面图

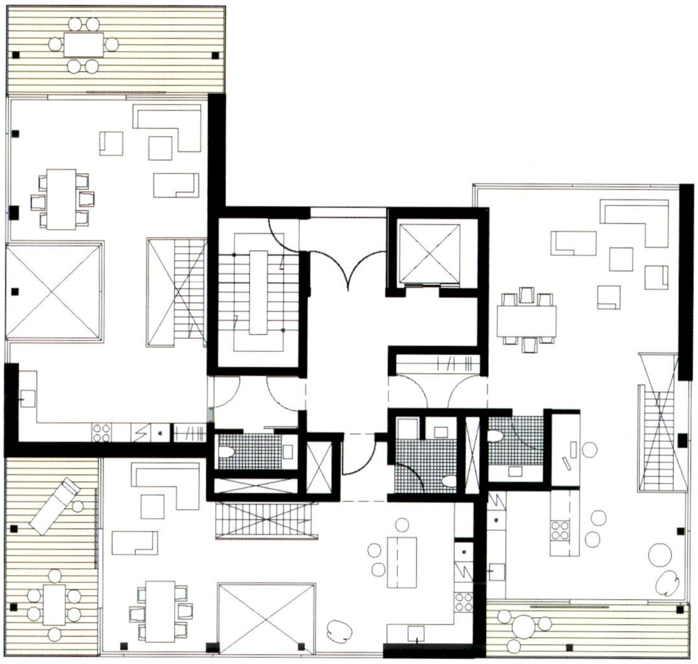

2nd Floor Plan　2层平面图

1st Floor plan　1层平面图

Social Housing for Mine-workers, Asturias, Spain

Location: Degaña, Asturias, Spain
Architect: ZON-E Architects
Gross Floor Area: 2,385 m^2
Number of Floors: 4
Photographer: Ignacio Martinez, Jose Antonio Ruiz

西班牙阿斯图里亚斯省矿工的社会福利房

地点：西班牙阿斯图里亚斯省黛格尼市
设计公司：ZON-E 建筑事务所
建筑面积：2 385 m^2
层数：4
摄影师：伊格纳西奥·马丁内斯、何塞·安东尼欧·鲁伊斯

Surmount 突破点

SKIN 表皮

SHAPE 造型

BAY WINDOW 凸窗

ENERGY SAVING 节能

This project comes up from a tendering process to build state subsidized housing in Cerredo (Asturias), a mining town located in the very heart of the Cantabrian Mountains where no residential construction had been made for over 25 years.

The project has two stages that materialize in two perpendicular buildings forming an L. In the first stage the designers undertake the biggest building, which faces the road that crosses the town.

The formal result is something halfway between a petrified object, a mountain's shape and a disturbing organism floating over the mountainside. Like a piece of coal, this "crystallographic" object absorbs almost all the light it gets and reflects a small amount of it. The skin of this "crystallographic" object has the same dark color as the local slate, calmly showing us its rich geometry.

The building's unity contrasts with the individuality of each of the 15 apartments that show through some galleries in the façade. These are cubes which drill the volume using a herringbone pattern and work as heat and light exchangers.

Each of the apartments is different, both in size and in its floor plan distribution, in the location of its gallery and in its roof's configuration. However, all of them enjoy cross ventilation and breathtaking views of Asturias' craggy landscape.

该项目的建造得益于政府要在阿斯图里亚斯的 Cerredo 建设国家保障型住房的招标。Cerredo 是位于坎塔布连山脉中心的一个采矿小镇，已经超过 25 年没有进行过居民住宅建设。

该项目分两个阶段，由两栋相互垂直的建筑构成，呈“L”形分布。在第一阶段，设计师主要负责完成最大的那栋建筑，面向穿过小镇的大路而立。

整栋建筑看起来就像是一个漂浮在山坡上的半石化、半山体状的奇异有机体，就像煤一样几乎吸收了所有照射到它身上的光线，只产生了少量反射。和当地的板岩相同，这栋“晶体外形的几何体”的建筑以暗色调呈现，平静地展示了其丰富的几何结构。

建筑的整体外形与通过立面突出的门廊展示出来的 15 间公寓形成鲜明的对比。这些门廊呈立方体，以“人”字形排列在体量上，用作内外热量与光照的交换器。

每间公寓的面积、平面布局、门廊位置，以及屋顶结构都各不相同。但是，所有房间都通风良好，能欣赏到阿斯图里亚斯壮丽的怪石景观。

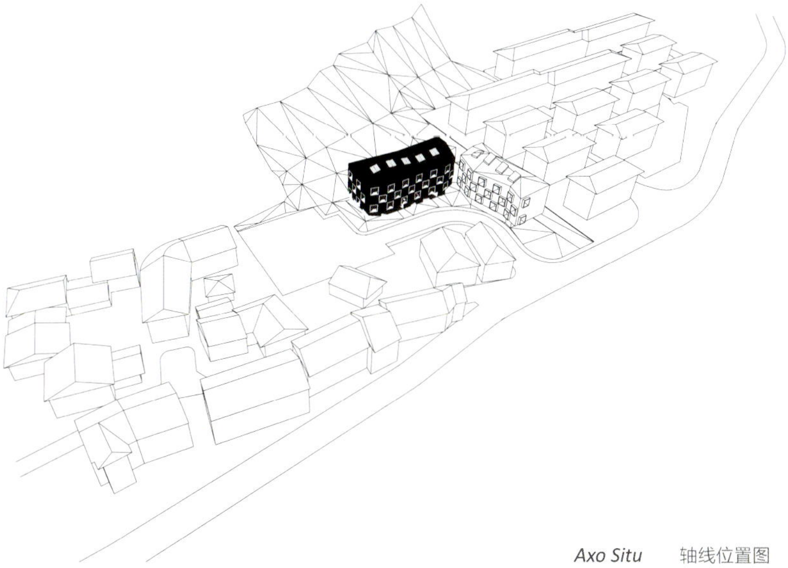

Axo Situ　轴线位置图

Site Plan　总平面图

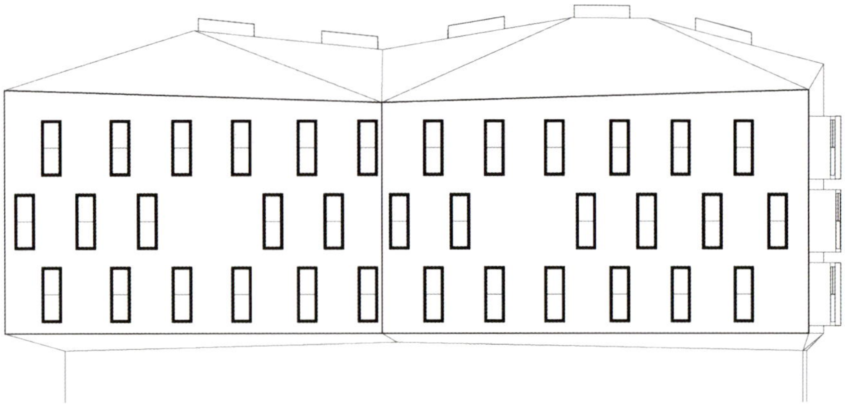

North Elevation Plan 北立面图

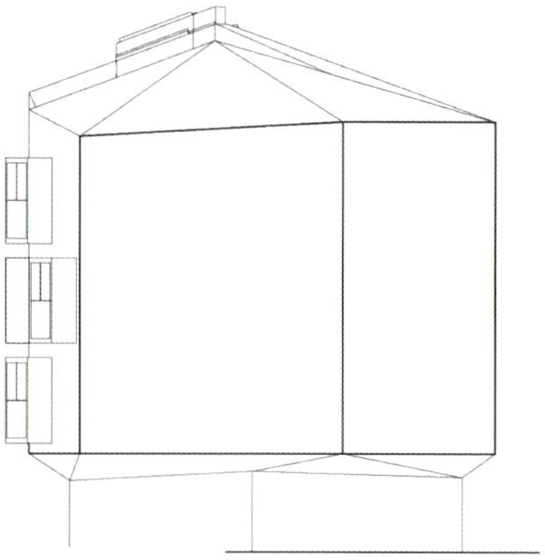

East Elevation Plan 东立面图

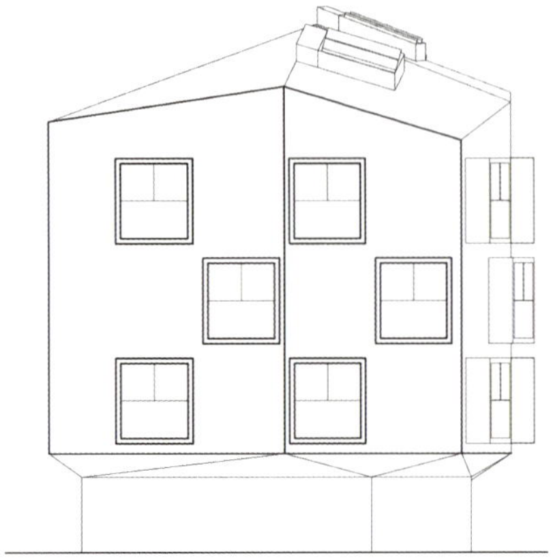

West Elevation Plan 西立面图

South Elevation Plan 南立面图

FEATURE 特点分析

Shape

According to the regulation of the town plan, the project is designed as angular shape like geometry crystallized from and contracted at the ground floor to emphasize the architect. The corridor makes the building more livable.

造型

由于小镇规划的一些基本条例，它的外形被设计成类似矿物晶体的多角形，并在底层收缩以强调建筑的物性，立面的门廊则把该建筑变得更宜居。

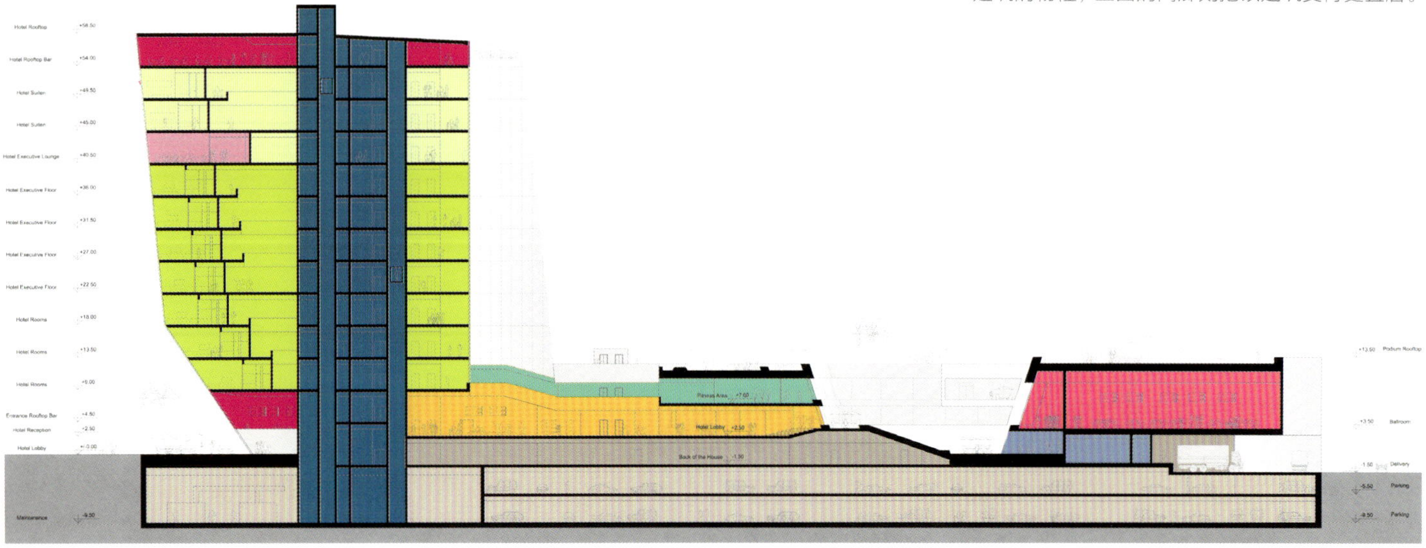

Section A-A A–A 剖面图

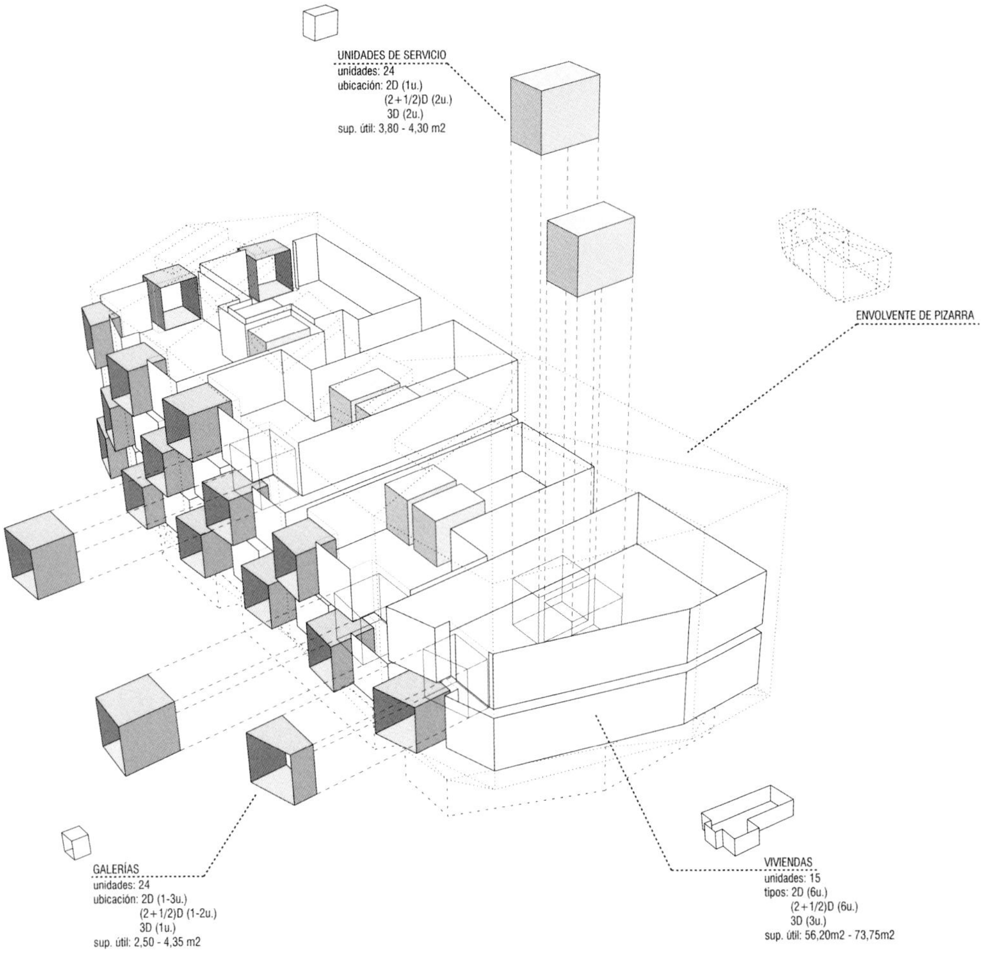

Axonometric Drawing 轴测图

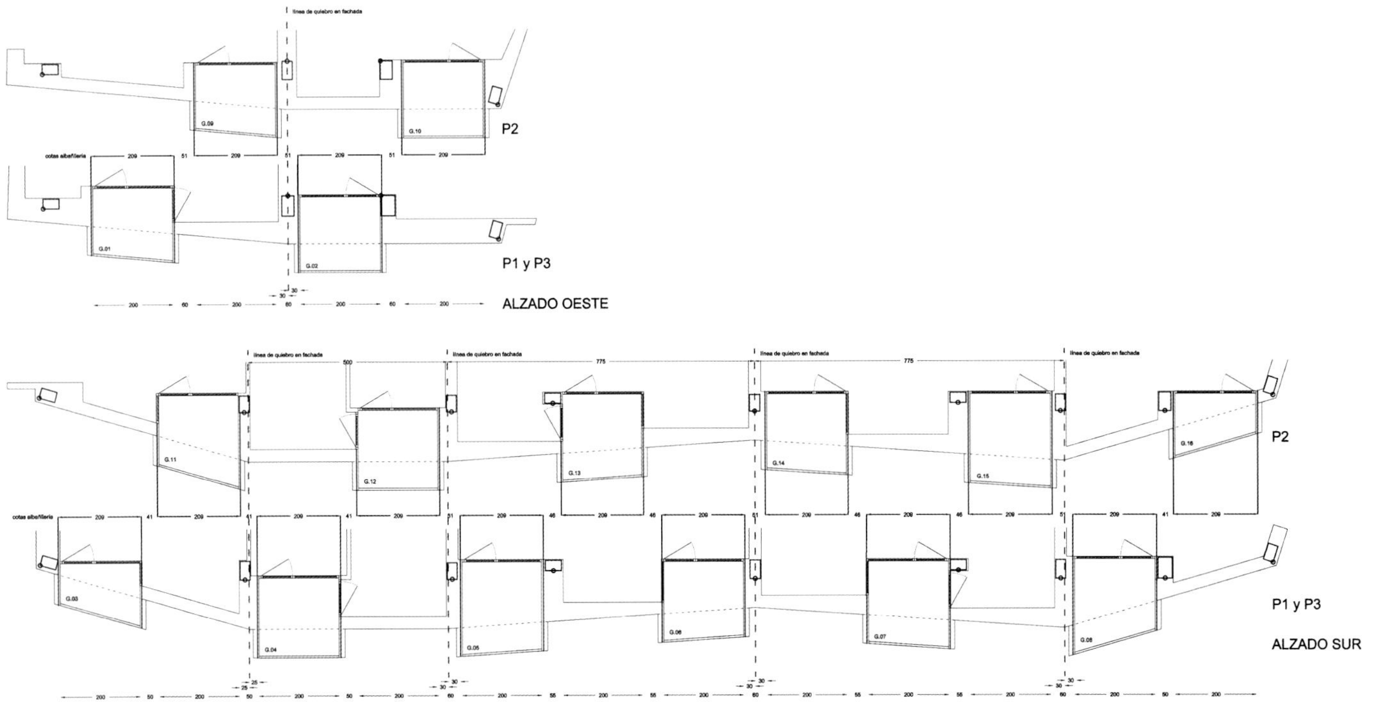

Detail Plan 细节图

Kajkanten, Copenhagen, Denmark

Location: Islands Brygge, Copenhagen
Architect: Danielsen Architecture
Gross Floor Area: 6,000 m^2
Number of Floors: 8

丹麦哥本哈根 Kajkanten 综合公寓楼

地点：哥本哈根卑尔根岛
设计公司：丹尼尔森建筑事务所
建筑面积：6 000 m^2
层数：8

Surmount 突破点

SCALE 尺度

SKIN 表皮

BALCONY 阳台

MATERIAL 材料

Danielsen Architecture have formulated a new housing development along Island's Brygge's prominent foreshore. Currently under construction, the 62 apartment complex "Kajkanten", stands 8 storeys high, topped with penthouses, and grosses a total floor area of 6,000 m^2.

"Kajkanten" is situated on the divide between Copenhagen's calm, blue harbor waters and the lush, green landscape of the adjacent urban landscape. In an architecturally diverse and expressive environment, surrounding residential compositions have further highlighted the prominent location of the project and the need to formulate a sensuous design.

Subsequently the development work and construction management has been innovative and inspiring, constantly striving to perfectly place "Kajkanten" within the new identity of Island's Brygge. "Kajkanten" will form a visual connection with Havneholmen on the opposite side of the harbor, whilst simultaneously anchoring itself architecturally in Island's Brygge.

The apartments maintain surrounding property heights and responds respectfully to the scale of the area.

丹尼尔森建筑事务所在卑尔根岛著名的前滩制定了一个新的建筑发展计划。这座建设中的 Kajkanten 综合公寓楼共有 62 套房间，8 层高，顶层是空中别墅，总建筑面积达 6 000 m^2。

哥本哈根港口水域平静而幽蓝，相邻的城市景观葱葱郁郁，Kajkanten 公寓楼就坐落于两者之间。建筑的多样化和富有表现力的环境，与周围的住宅楼群形成对比，进一步突显了项目优越的地理位置和感性设计的必要性。

随后的项目工作和建设管理不仅具有创新性，也非常鼓舞人心，使 Kajkanten 综合公寓楼完美地成为了卑尔根岛的新地标，与卑尔根岛和谐地融合在一起。

公寓楼和周围的楼群都保持同等的高度，也和该地区的规模相呼应。

FEATURE 特点分析

Skin

The façade is clad white plastered and speaks elegantly with the copper-like material tombac that runs along the balcony edge and floor of the penthouses above. The visual connection with Havneholmen on the opposite of the harbor is achieved through functional solutions and the use of traditional surfaces that refer to rich Scandinavian architectural foundations.

表皮

立面采用白色石膏覆层，阳台的边缘和空中别墅的地板则采用了优雅的铜质顿巴合金。在斯堪的纳维亚深厚的建筑基础上，公寓楼结合了建筑的功能性和传统表皮的使用，与港口对面的 Havneholmen 形成视觉呼应。

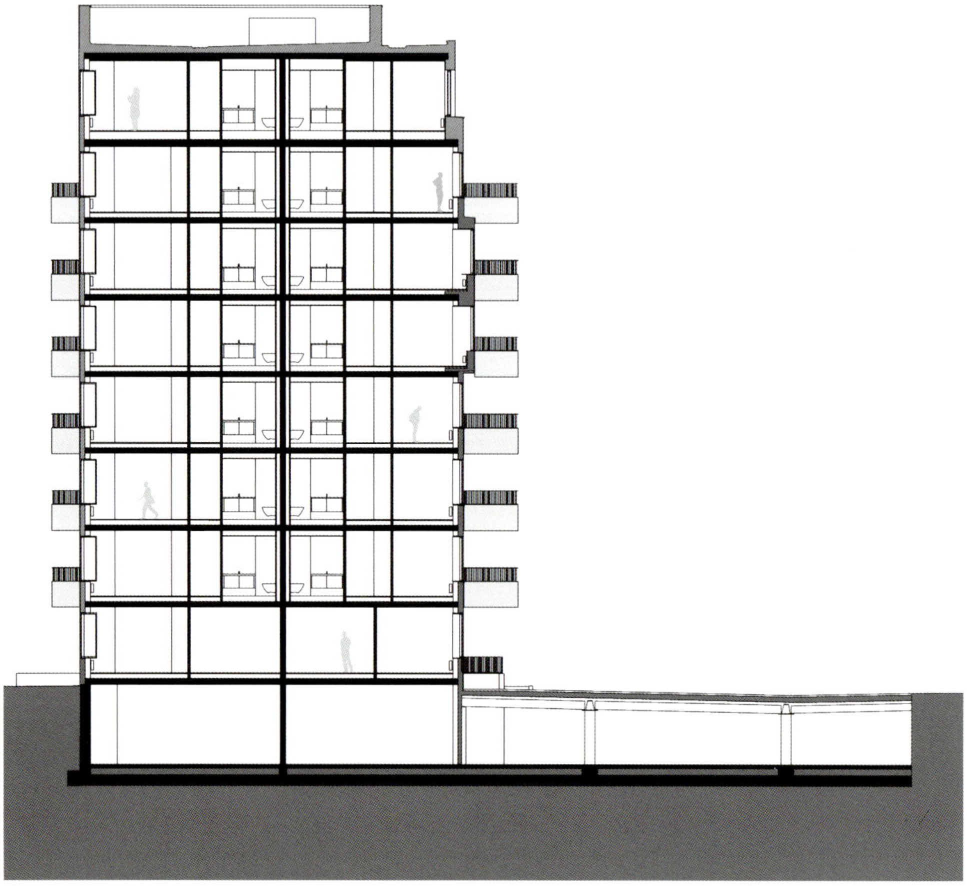

Cross Section Plan 横截面图

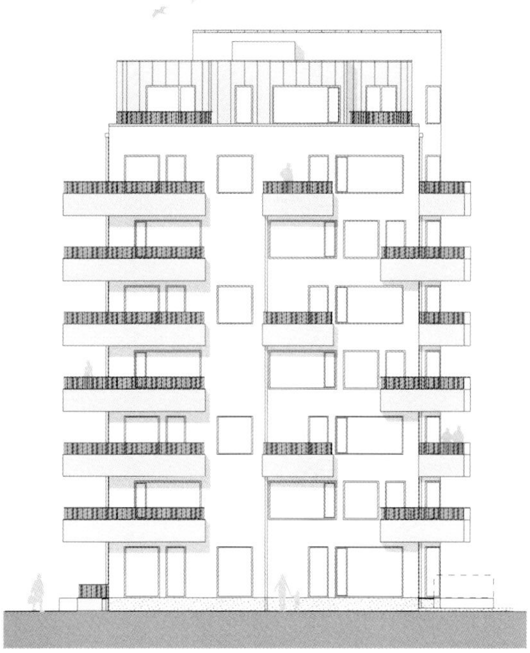

West Elevation Plan 西立面图

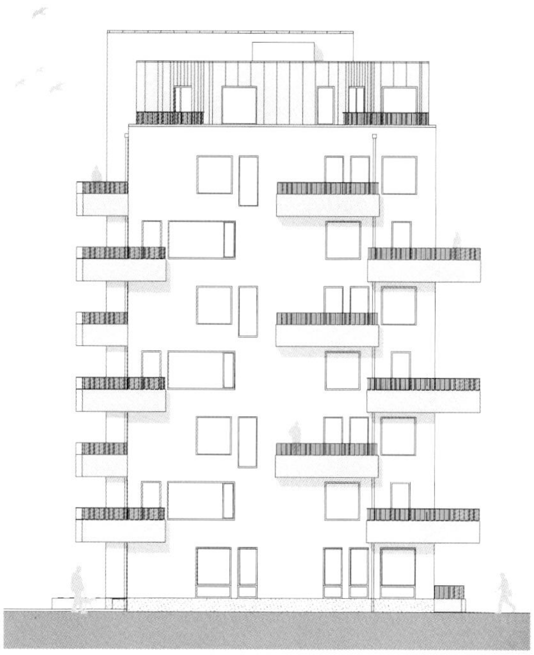

East Elevation Plan 东立面图

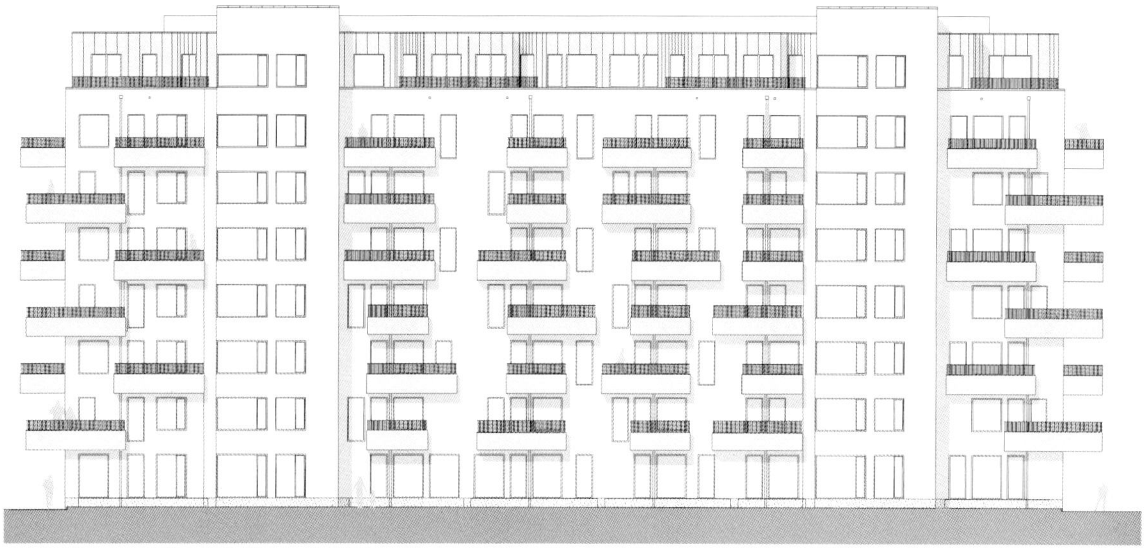

South Elevation Plan 南立面图

North Elevation Plan 北立面图

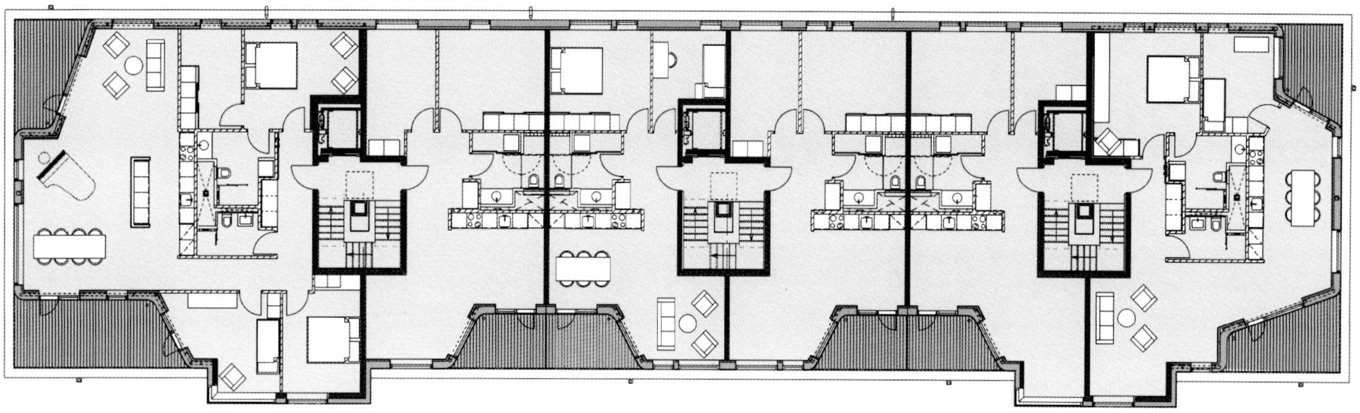

7th Floor Plan　　7 层平面图

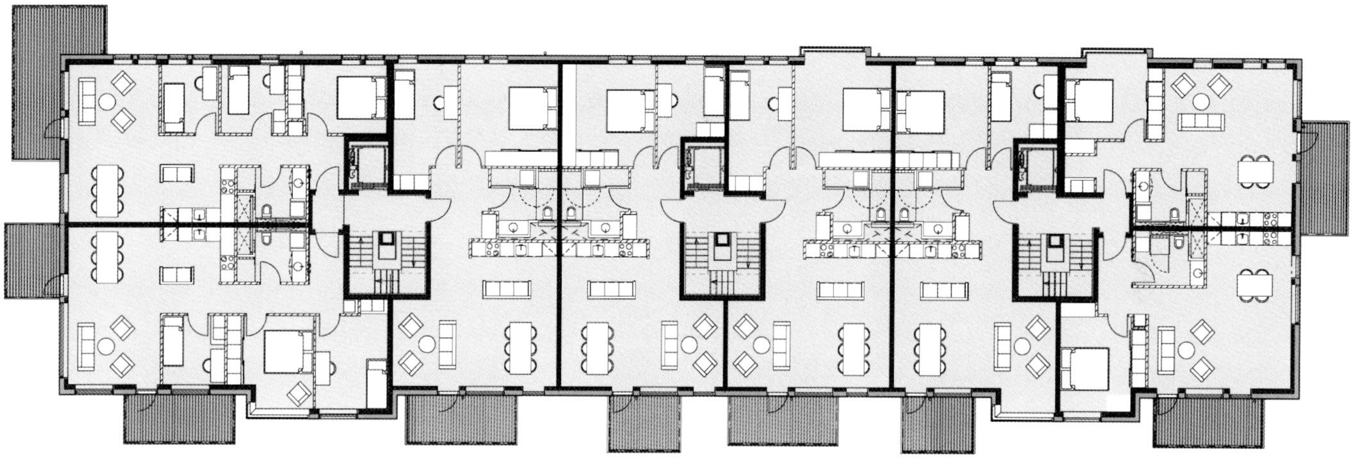

1st-6th Floor Plan　　1~6 层平面图

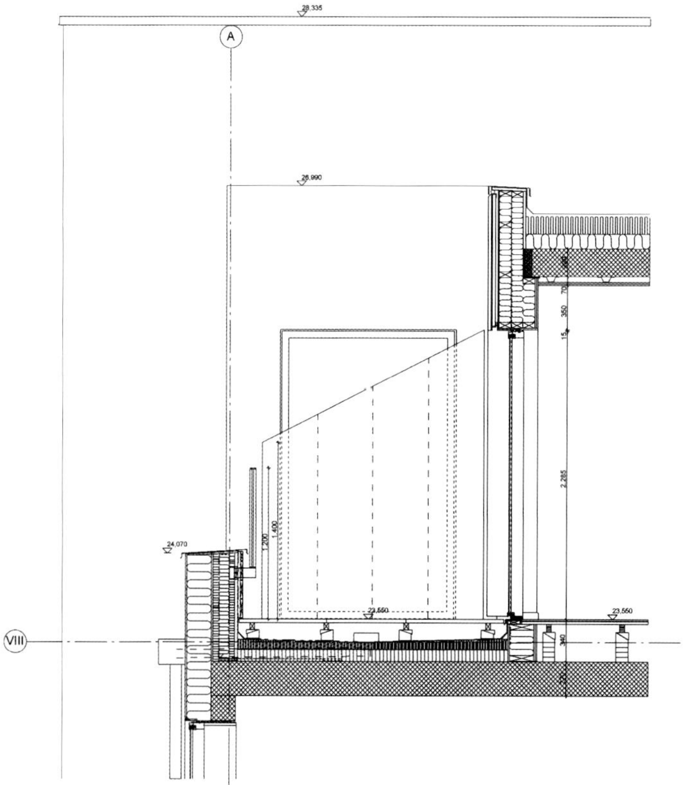

Section Detail Plan 1　　剖面细节图 1

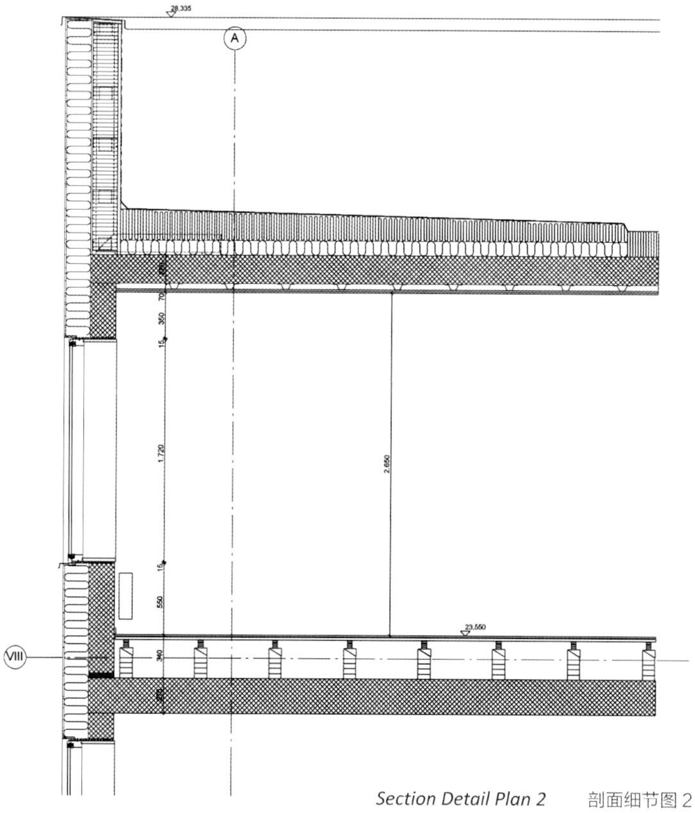

Section Detail Plan 2　　剖面细节图 2

Sonnenhof, Jena, Germany

Location: Jena, Germany
Architect: J. MAYER H. Architects
Gross Floor Area: 10,000 m^2
Number of Floors: 7

德国耶拿市赛纳霍夫办公住宅楼

地点：德国耶拿市
设计公司：J · 迈耶 · H 建筑事务所
建筑面积：10 000 m^2
层数：7

Surmount 突破点

SHAPE 造型

FACADE 立面

FASHION 时尚

MATERIAL 材料

Sonnenhof consists of 4 new buildings with office and residential spaces. Located on a consolidated number of smaller lots in the historical center of Jena, Germany, the separate structures allow for free access through the grounds. Their placement on the outer edges of the plot defines a small-scale outdoor space congruent with the medieval city structure. Its outdoor facilities continue the building's overall design concept past the edges of the lot. The planned incorporation of commerce, residence and office enables a small-sectioned and flexible pattern of use that also integrates itself conceptually into the surroundings.

Designed by J. MAYER H. Architects the four-building complex leaves a large part of the space open for public use during the day, permitting a free flow of pedestrian traffic across the area. The mixed-use concept supports a small scale, seamless integration into the existing urban fabric, integrating such functional elements as flower beds, ventilation openings, seating and lighting fixtures and thus transforming the open areas into attractive, urban leisure spaces.

赛纳霍夫办公住宅楼包括 4 栋带办公区与住宅区的新楼，坐落在德国耶拿市的历史中心地带，独立分离的结构使人们可以在首层自由地出入于各栋大楼。这 4 座大楼建造在场地边缘地区，中间围成一个小型的户外空间，类似于中世纪的城市布局。户外设施秉承了建筑整体的设计理念，沿着场地的边缘布置。集商业、住宅和办公于一体的多功能设计，赋予了建筑一种细分、灵活的使用模式，还把建筑本身完美地融入到周围的城市环境中。

在 J · 迈耶 · H 建筑事务所的精心设计下，这个 4 栋综合大楼为公众保留了一片很大的空间，方便行人在整个区域自由穿行。多功能的设计理念使建筑天衣无缝地融入到城市环境之中，与花床、通风口、座位以及灯具等一系列功能设施融为一体，把开放区域变成了极富吸引力的城市休闲空间。

1st Floor Plan 1层平面图 Scale 1 : 500

FEATURE 特点分析

Façade

The design of the outer buildings takes up the geometry of the polygonal shapes in the façade, continuing their graphics across the floor. The resulting surfaces are often extended and "programmed" into the third dimension, forming a sculptural set of wedges and steles.

立面

外部设计沿用平面图式，采用了多边形几何结构设计作为建筑立面，并把这种多边形的几何图案一直延伸到地面。这样的设计不仅延伸了建筑的表面，还给人以强烈的三维立体感，形成一系列雕塑般的楔形和柱状结构。

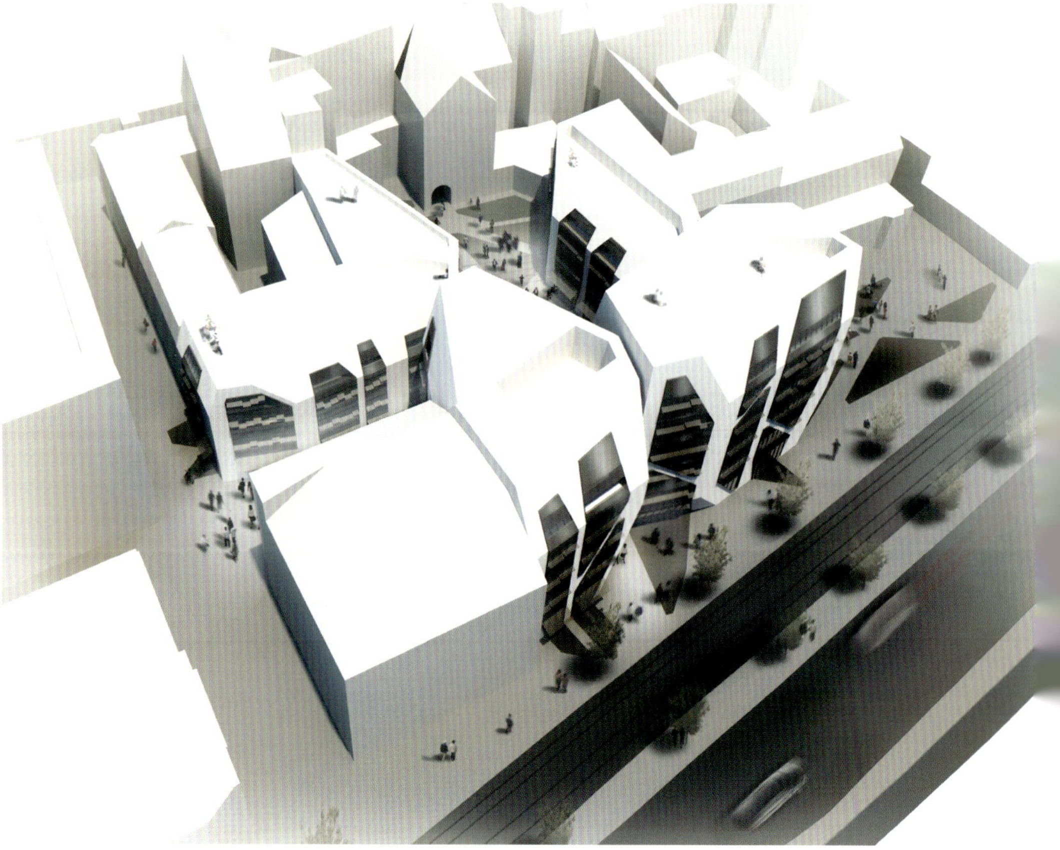

Model Plan 模型图

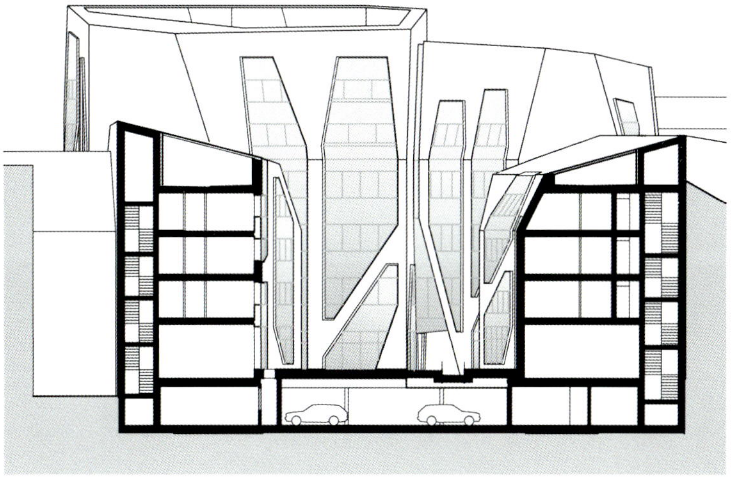

Section Plan A-A 剖面图 A-A

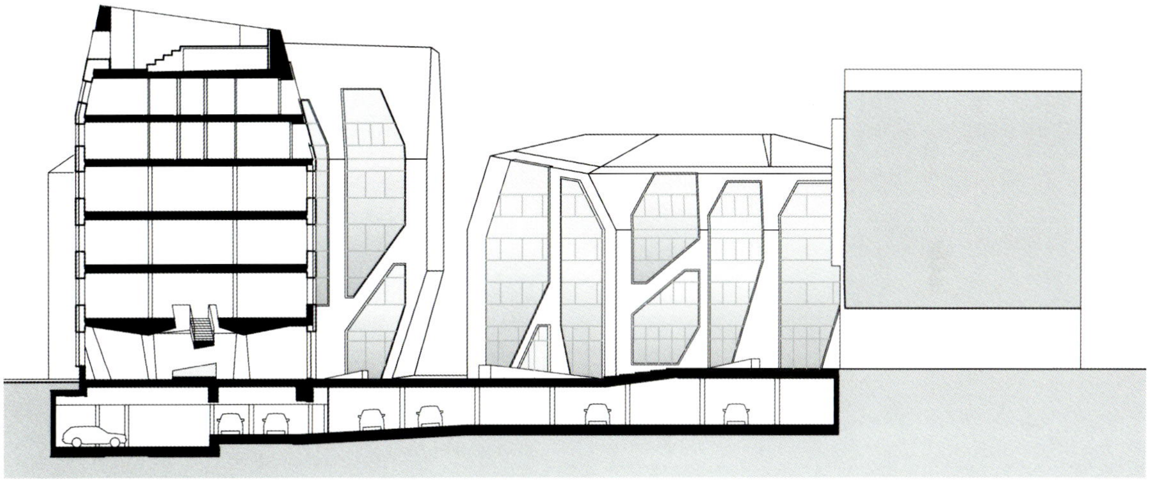

Section Plan B-B 剖面图 B-B

Strandkanten- K9 Housing, Norway

Location: Tromsø, Norway
Architect: 70°N Arkitektur
Site Area: Approximately 2,100 m^2
Number of Floors: 5
Photographer: Ivan Brodey, Magdalena Haggarde

挪威 Strandkanten 住宅区 K9 公寓楼

地点：挪威特罗姆瑟市
设计公司：70°N 建筑事务所
占地面积：约 2 100 m^2
层数：5
摄影师：伊凡 · 布罗迪、玛格达蕾娜 · 哈加德

Surmount 突破点

FACADE 立面
SHAPE 造型
ENVIRONMENT 环境
MATERIAL 材料

This project is part of the Strandkanten housing development along the eastern side of the Tromsø island. The project sits in the southern end of the area with the view to the south open to the mountains and the fjord.

The programme is of about 2,100 m^2 and consists of 23 apartments of varying size, all of which are designed with wheelchair accessability in mind. On the ground floor and exposed to the quai there is a small commercial area intended for a local cafe or similar function for the township. From the building there is also direct access to public outdoor areas on the first floor.

This volume is shaped with cutaways, one for the local pedestrian main street and one large cutaway to allow for terraced balconies to the south.

本案是沿着特罗姆瑟东边而建的Strandkanten住宅开发项目的一部分。该项目坐落于这片地区最南端的位置，可以将南面的山脉和峡湾一览无遗。

该项目约2 100 m²，共包括23间大小不同的公寓，特地配置有轮椅专用道。首层面向码头，有一个小型商务空间，里面有咖啡厅或类似这种为城区服务的商店。建筑首层还有通道直达公共区。

这个体量被设计成斜切型，一部分作为主人行道，另一部分较大的斜切面则变成了一个向南的分层的露台。

Location Plan 区位示意图

FEATURE 特点分析

Shape

Slope and skew of the volume relates to limit of the site, while it does break the modern architecture's traditional box shape.

造型

建筑造型的斜切，呼应场地的环境，也打破了"方盒子"的现代传统。

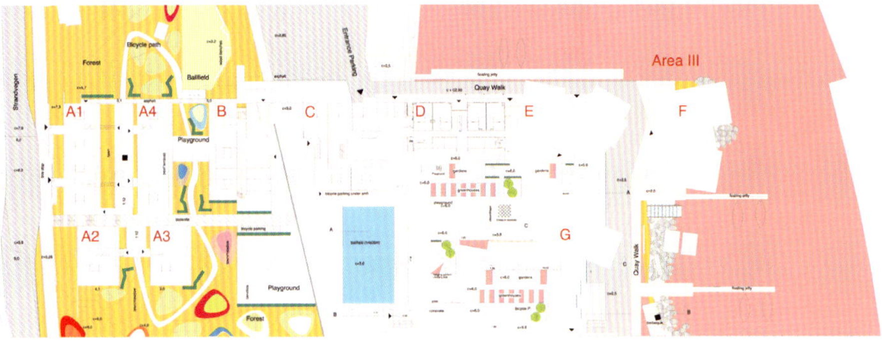

Site Plan 总平面图

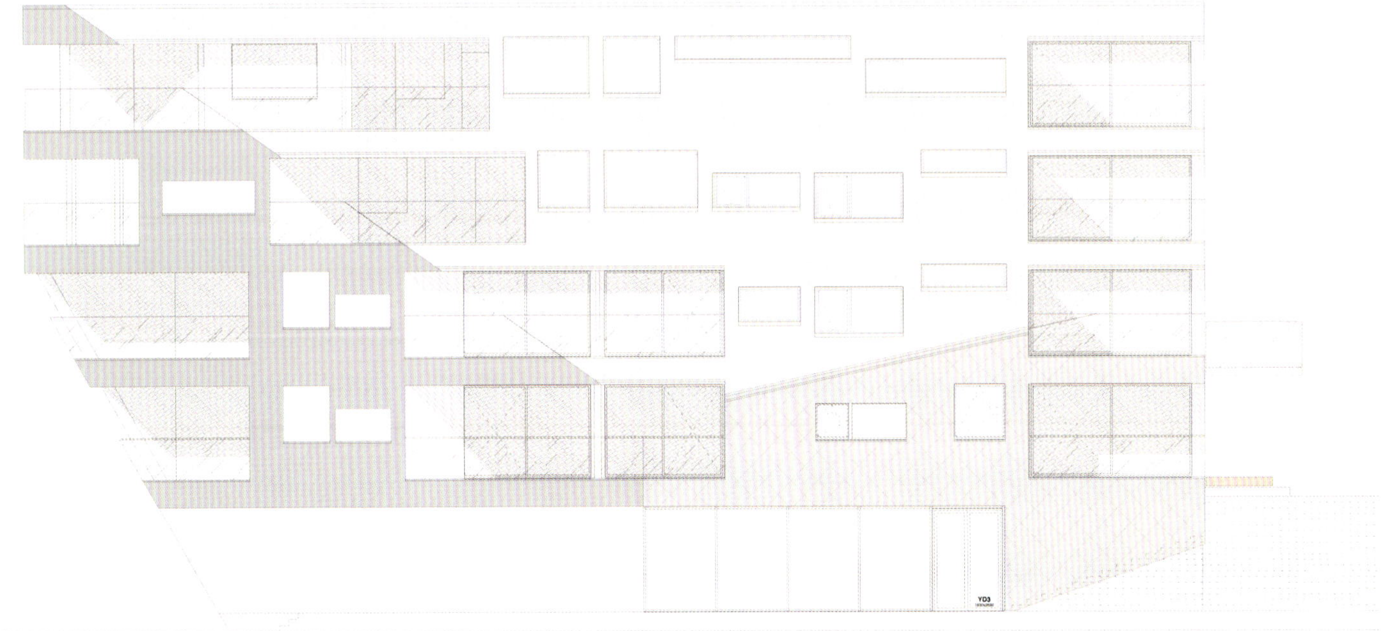

Elevation Plan 1 立面图 1

Elevation Plan 2 立面图 2

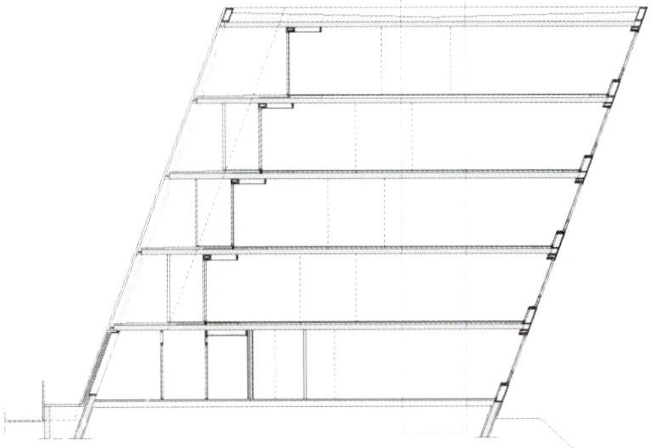

Section Plan 剖面图

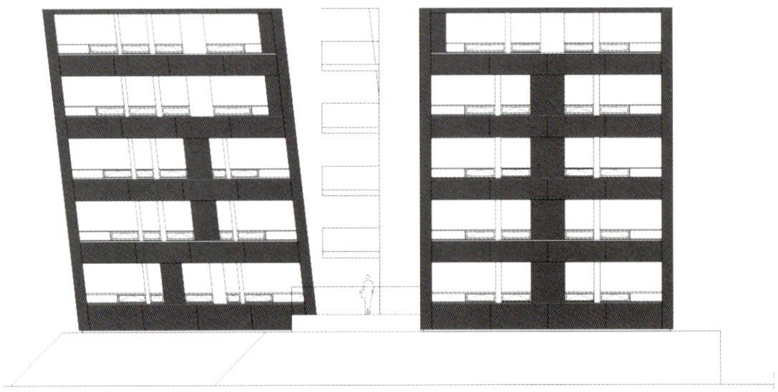

Elevation Plan 立面图

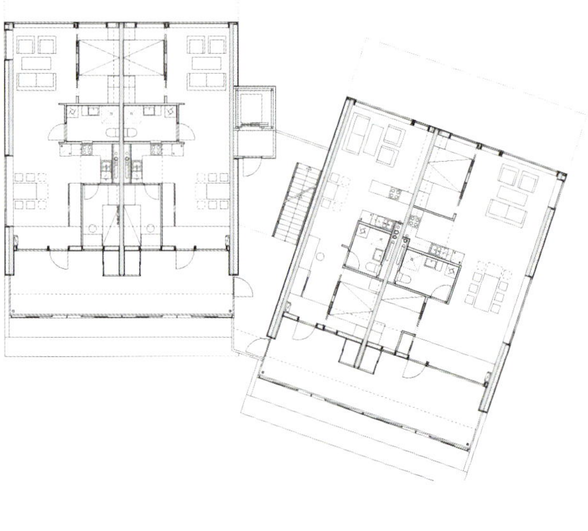

Floor Plan 楼层平面图

Floor Plan 楼层平面图

Blok K, Amsterdam, The Netherlands

Location: Amsterdam, The Netherlands
Architect: NL Architects
Gross Floor Area: 1,815m²
Number of Floors: 2.5
Photographer: Luuk Kramer, Raoul Kramer, NL Architects & Roel Backaert

荷兰阿姆斯特丹市 Blok K 住宅楼

地点：荷兰阿姆斯特丹
设计公司：NL 建筑事务所
建筑面积：1 815 m²
层数：2.5
摄影师：Luuk Kramer, Raoul Kramer, NL Architects & Roel Backaert

Surmount 突破点

SHAPE 造型

ROOF 屋顶

LAYOUT 布局

GREEN 绿色

Blok K is located in a triangular site between the historic center and the recently redeveloped harbor area in the east of Amsterdam. Along the east and the south side a "wall" containing over 300 apartments and office spaces shields the site from the noise of the adjacent railroad. Inside this semi open block a loose grid is set up, containing 16 smaller housing blocks positioned in a Park. These "Hidden Delights" vary in height from 9 m to 18 m.

Blok K is one of 3 almost square blocks that measure 31 m by 28 m and should contain 2.5 storeys. It was obligatory to build the first two stories in alignment; the third should be 50% building and 50% roof terrace/garden. In our block the volume is distributed evenly over the ten houses; each is allocated 630 m³.

The houses are organized according to the back to back principle. They will be accessed from an Aisle in the middle of the block. This "mini-canyon" rids the façades of the obligatory storage spaces and technical facilities that formally have to be accessible from the public domain. The front doors are positioned in the back; the houses are turned inside out. By placing the "intestines" like water and gas meters, hallways and stairs in the darker zones of the apartments, the façades can open up to the light and to the view, to where the park is.

Three other buildings with a slightly smaller footprint, but considerably higher (18 m), rise up next to the block. By interpreting the given 2.5 floors as 2 + ½ story as in Being John Malkovich we could make a "clean" start without formal determination; a block of 6 + 1.5 = 7.5 meters high.

To create space within the master plan this volume is deformed and redistributed, pushed away as much as possible from the adjacent volumes. Within the orthogonal grid a diagonal vista opens up. The design tool deployed for this operation is the slight rotation of the Aisle, orienting it onto two open spaces between the blocks, instead of two "blind" walls and as such providing a public shortcut.

Point of reference is the given 630 m³ per dwelling. While all houses are at the same time stretched and compressed, either in height or in length, they still maintain this original volume. Although some features of the typology remain, all houses are individual and unique reaching from 1.5 to 4 storeys, from 132 m^2 to 185 m^2. All houses are different but part of a larger whole; they are one. The amplitude of the building now varies from 5 m to 15 m, but overall the average is still the required 7.5 m. In the process the standard building bay as an organizational tool of construction became "elastic".

Blok K 项目位于一块三角形的场地上，处于东阿姆斯特丹的历史中心与最新重建的海港区之间。沿东南侧的 300 间住房和办公空间形成了一道“墙”一般的屏障，隔绝了来自邻近铁路的噪声。这个半开放公园街区内部采用了松散的网格布局，包括 16 栋大楼，高度从 9 m 到 18 m 不等， 如同“深藏不露的风景”。

Blok K 住宅是 3 栋长 31 m、宽 28 m 的准正方形建筑的其中之一，共 2.5 层。该住宅的第 1、2 层面积一样，而第 3 层只占一半的面积，另一半用作露台或花园。里面的 10 间住房大小相同，每间都有 630 m^3。

住宅楼的居住单元按照“背靠背”的原则布置，由大厦中间的一条过道联结着。这条“小峡谷”从通常需要从公共领域进入的存储空间和技术设施的立面中独立出来。单元的前门位于后面，它“里外颠倒”了。通过把水表、气表、过道和楼梯这些设施放在较暗的位置，外立面得以展现在阳光与风景中，朝向公园。

其他 3 栋建筑的占地面积略小，但是相对都比较高（18 m），紧靠着住宅楼。设计师把 2.5 层分解成 2+$^1/_2$ 层，就像在电影《傀儡人生》里一样，不需要正式的测定便能精准地开工，建造出一栋 6+1.5=7.5 m 高的建筑。

为了在总体规划中创造出更多的空间，建筑的体量需要进行变形和再分配，尽可能与相邻建筑拉开距离。于是设计师在正交网格里开辟了对角视野。利用走道位置的轻微旋转，使它朝向建筑之间的两个开放空间，代替了原本两堵阻挡视线的墙，变成一条公共捷径。

尽管所有的居住单元的长度、高度都经过了拉伸或压缩，每套单元仍然保持有 630 m^3 的空间，同时依旧保持着建筑的原始外形。此外，所有的房间在保留了特定类型的一些特征的同时，都极具个性且独一无二，分别从 1.5 至 4 层，132 m^2 至 185 m^2 不等。尽管房间都各不相同，却又组成一个统一的大整体。建筑的高度在 5 m~15 m 间，但整体还是保持在所要求的 7.5 m。项目的标准隔间这一组织工具变得具有弹性。

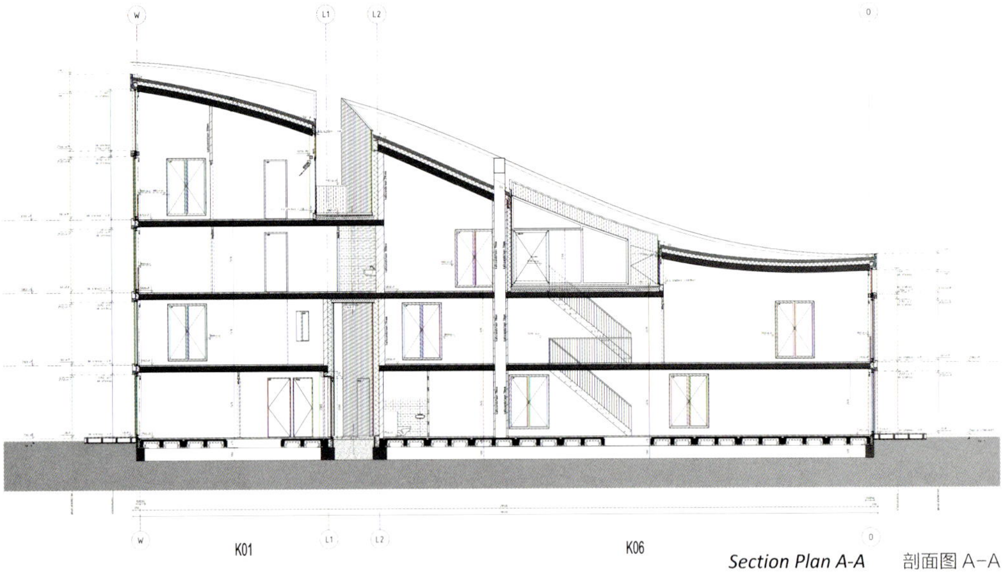

Section Plan A-A 剖面图 A-A

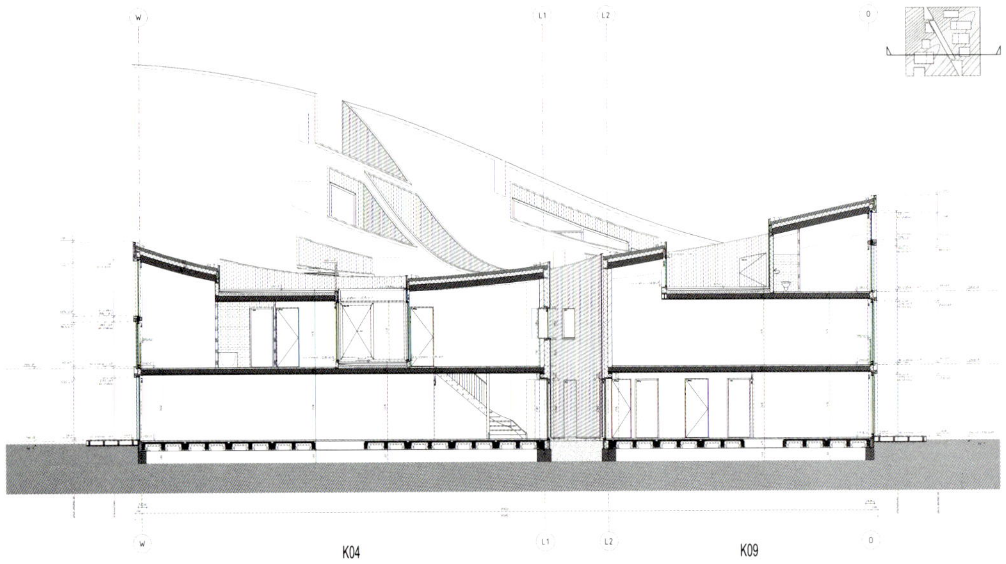

Section Plan D-D 剖面图 D-D

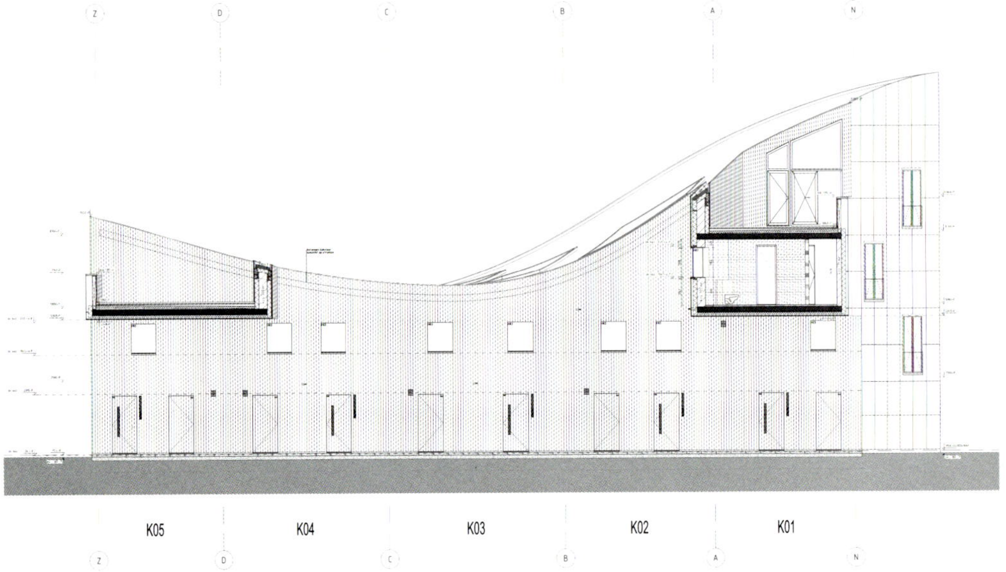

Section G-G 剖面图 G-G

FEATURE 特点分析

Roof

The sank round roof is en eco-roof skin, which will become a more open space. Roof terrace is embraced by fully greens with a central corridor to make the whole building lighted and well-ventilated.

屋顶

下沉曲面屋顶本身是一种生态表皮，也能成为更大的开放空间，屋顶阳台是拥抱绿意的室外空间，与中间的通道一起使整个建筑采光充分，通风良好。

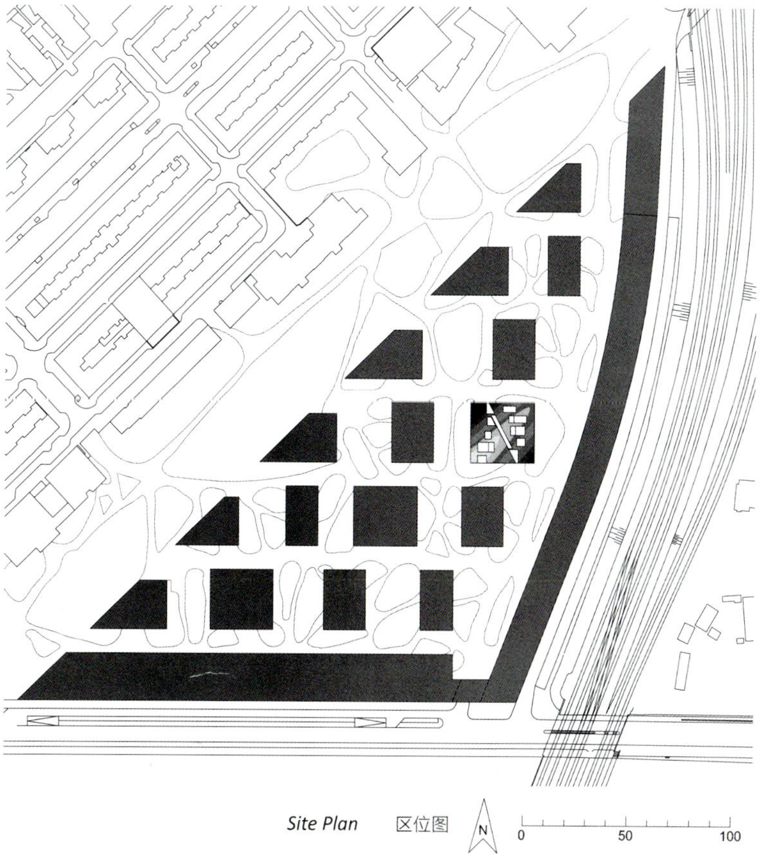

Site Plan 区位图

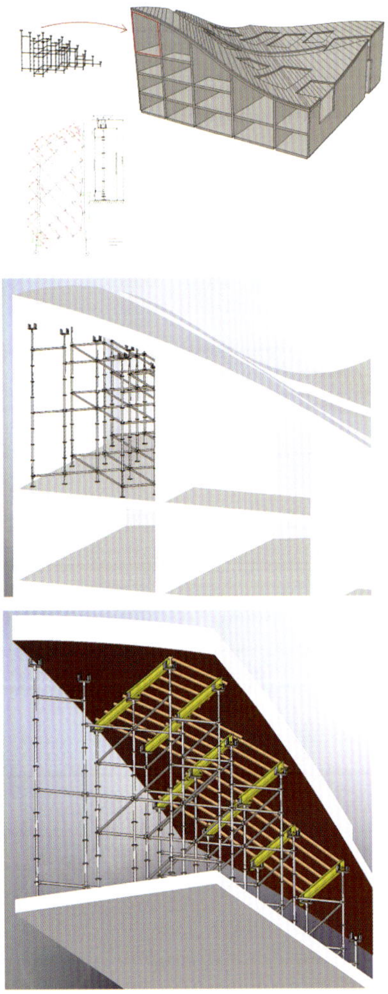

Structure Analysis Plan 结构分析图

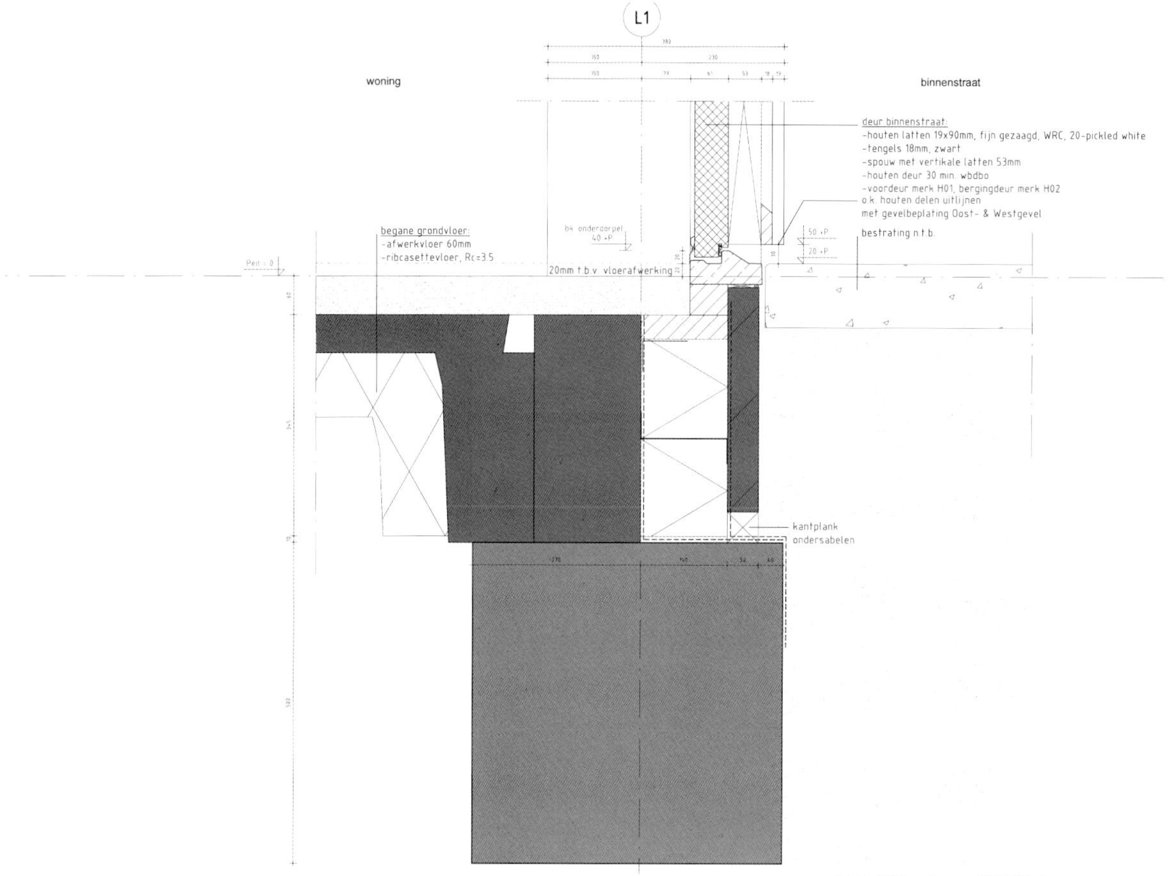

Detail Plan 1　　细节图 1

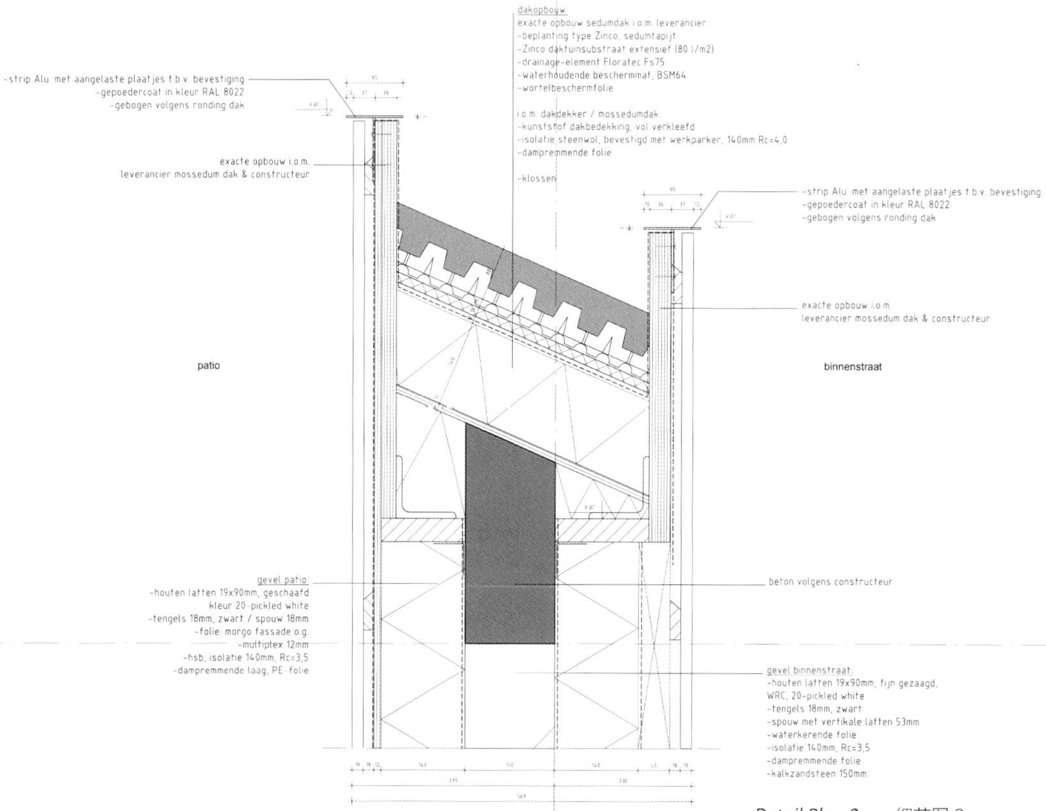

Detail Plan 2　　细节图 2

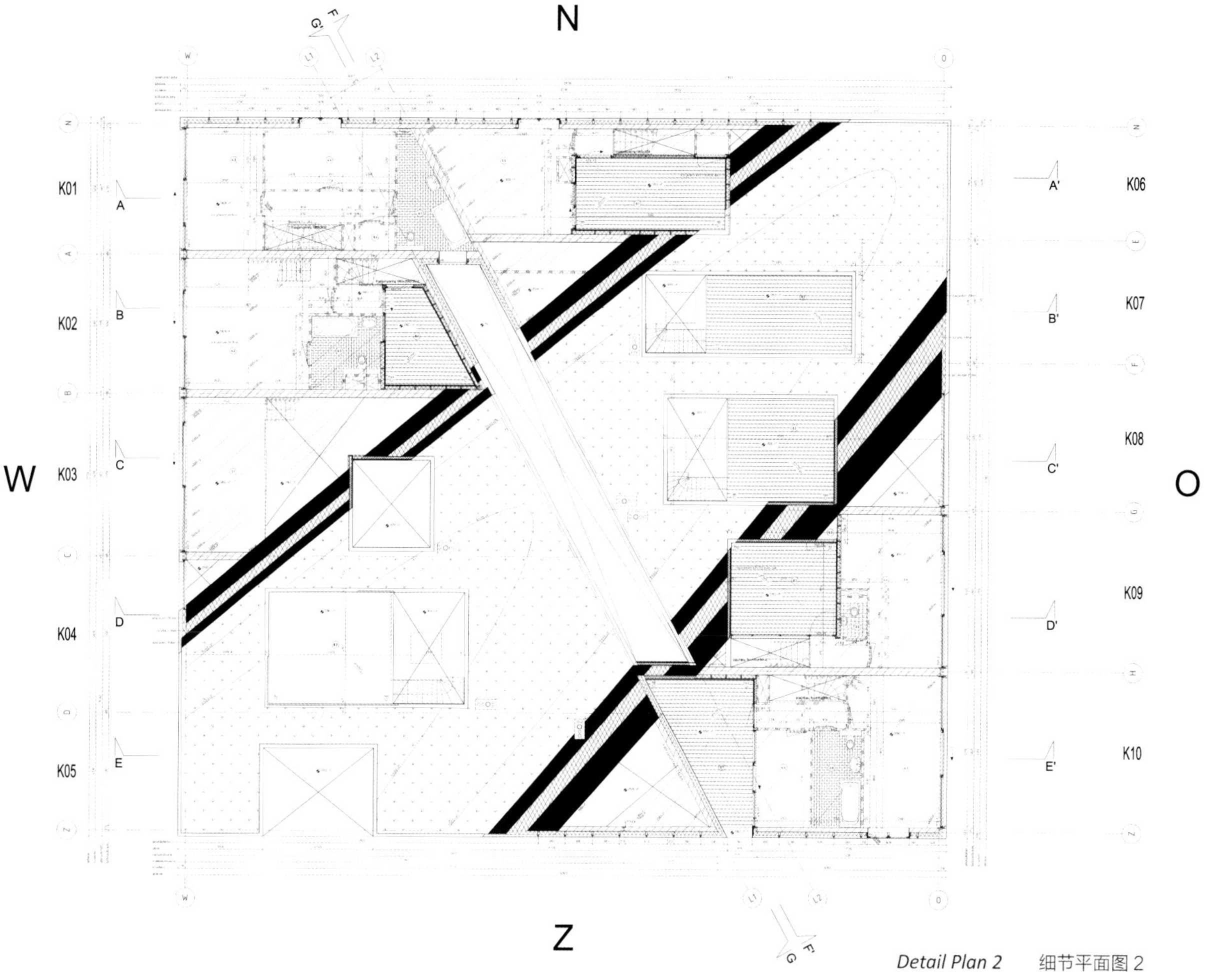

Detail Plan 2　细节平面图 2

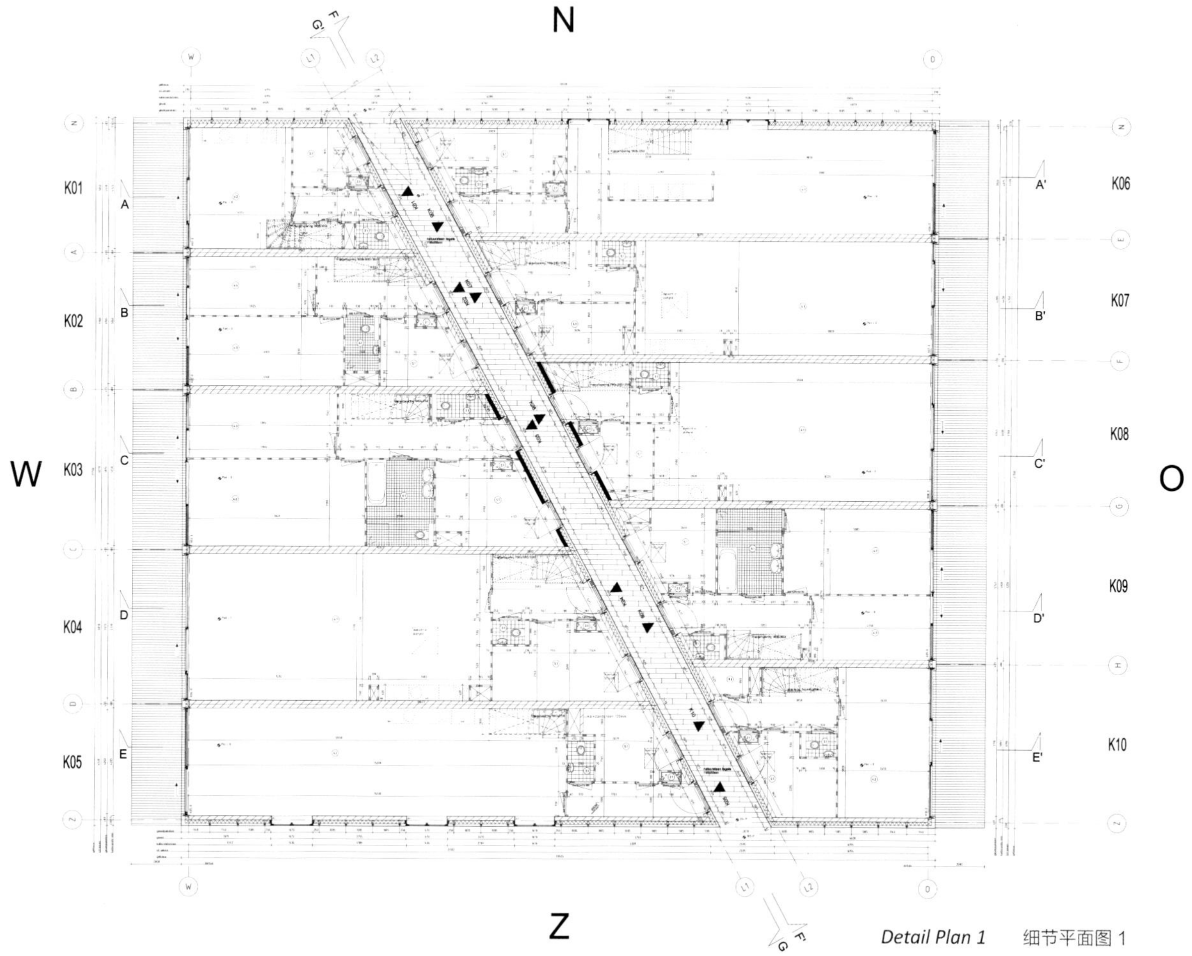

Detail Plan 1　细节平面图 1

Citadel, The Netherlands

Location: City of Westland, The Netherlands
Architect: Waterstudio.NL
Height: 12 m
Number of Floors: 3

荷兰 Citadel 城堡漂浮公寓

地点：荷兰韦斯特兰市
设计公司：水工作室
高度：12 m
层数：3

Surmount 突破点

LAYOUT 布局

SHAPE 造型

MODULAR 模块化

STRUCTURE 结构

In Holland in the city of Westland, Waterstudio is designing now a project that is contributing on a new relationship with the water. This project is not one of the livings next to the water but a living on water. The Citadel is Europe's first floating apartment building.

The citadel is part of an urban development with 1,200 houses that will be built in a depolderised zone, called the New Water. The New Water is a polder that will be deliberately filled with water after a few centuries artificially dry period. The depolderising is done to protect other surrounding polders for flood during times of heavy rainfall.

The toolbox of floating developments in Holland is still not so large, floating houses, villas are normal, but concepts with high density are not available. The Citadel consists of 60 units, which makes it the first floating development with more than 30 housing units/4,047 m^2 water. This relatively high density leaves more open water surrounding the building. The floating building is a composition of 180 modular elements around a courtyard on top of a floating concrete caisson foundation. All the apartments have waterview and most of them have a berthing place for a small boat. The Citadel will be projected on the middle of the water with a floating road for connection with the shore. The comfort level for this development is the same as in for instance a high rise building.

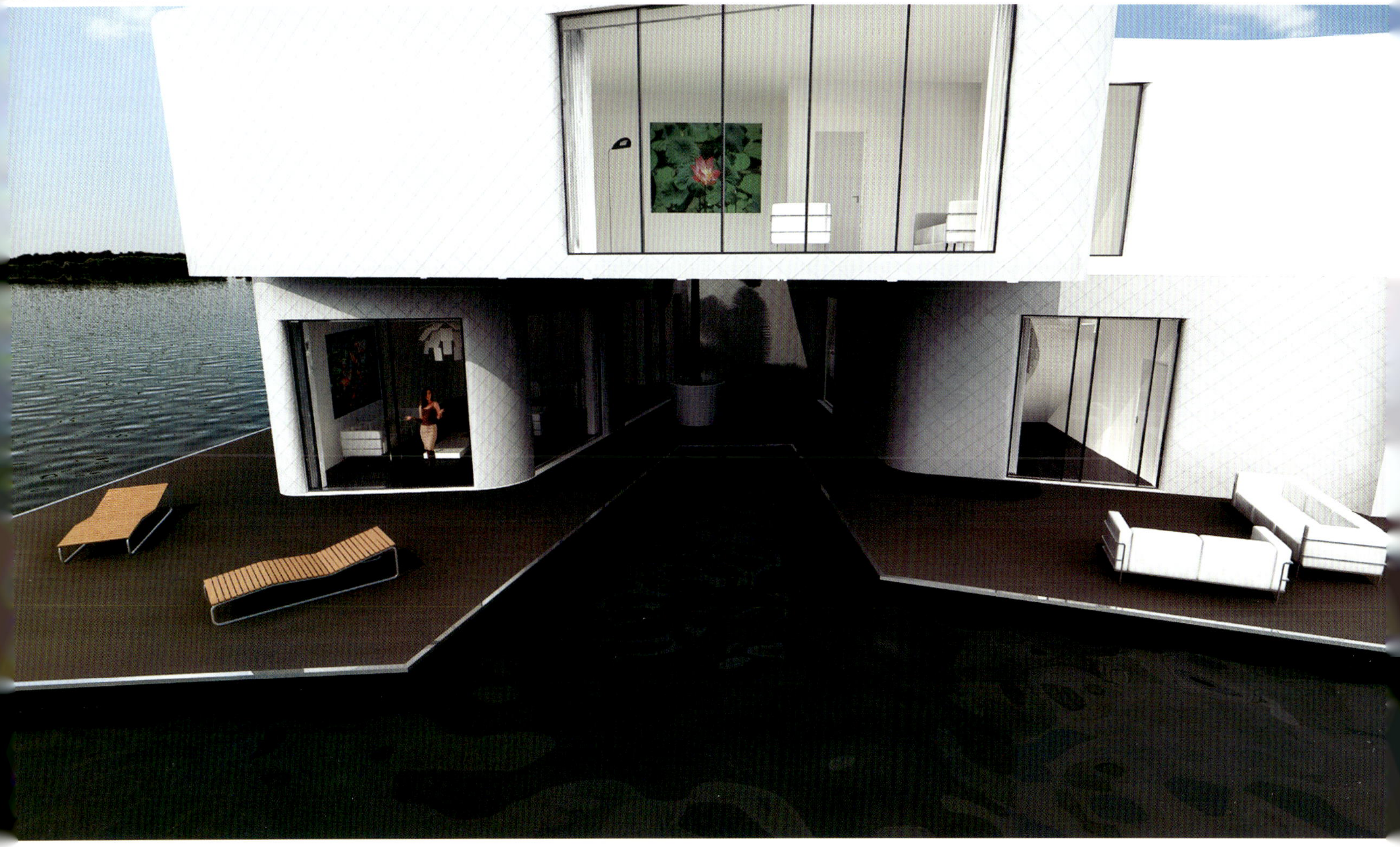

Sustainability is a focal point for the principal of the development. As first floating apartment building, the Citadel has the possibility to be a benchmark for floating urban development. The façade of the building is from aluminium panels. Those cost a bit more energy to produce but they have a very long lifespan and less maintenance on this water location. The energy consumption over the total lifespan will be 25% less than a conventional building on land.

In the New Water a mix of archetypes with different prizes is being developed. As a semi-governmental development, the project serves a social economical and ecological purpose. Each unit of the citadel has a different floor plan and outside space. The maisonettes in the apartment building open the possibility for a rich variety of double layered outdoor spaces. The construction of the houses is engineered in such a way every client can design his own personal floor plan within the determined shapes.

Ground Floor Plan 首层平面图

Top View Rendering 顶视效果图

在荷兰的韦斯特兰市，水工作室正在设计一个新型亲水项目。这个项目不是建在水边，而是浮在水面上。它是欧洲首个漂浮式公寓建筑。

Citadel 是城市发展项目计划的一部分，位于一个叫“新水”的还海区内，计划建造 1 200 栋房屋。经过了几个世纪的人工排水，水将会被有目的地灌注到“新水”还海区。还海区的建立是为了保护附近的开拓区在雨季来临时不被洪水淹没。

在荷兰，漂浮的小房屋和别墅很常见，但能漂浮的大项目却很罕见，而高密度的概念还从未实现过。这个由 60 个单元组成的城堡是首个密度能达到每 4 047 m^2 有 30 多个住宅单元的项目。这种高密度使建筑四周能有更多开阔的水面。项目建立在一个漂浮的混凝土沉箱地基之上，由包围着庭院的 180 个模块化元件组成。所有公寓都能看到海，而且大多数都有小艇停泊位。项目将建立在水中央，经由浮路通向陆地。这个建筑项目的舒适程度与一栋高层建筑的舒适水准相若。

可持续性是项目开发原则中的焦点。由于 Citade 是首个漂浮式公寓，因此它可能会成为城市漂浮项目的基准点。该项目的公寓楼四周被温室所包围。建筑立面由铝板组成，虽然导致了更多的能源消耗，却在这个临水的地方能延长使用寿命和减少维护工作。与陆地上的传统建筑相比，该项目在它的年限里将节约 25% 的能源。

在“新水”还海区正开发着许多不同建筑风格的获奖漂浮项目。作为半官方发展项目，它的目标是经济与生态的双赢。本项目的每个单元的布局和户外空间都不一样，跨层公寓让多种的双层室外空间成为可能。项目施工的设计方式让每个客户都可以在既定外形下设计他的个性布局。

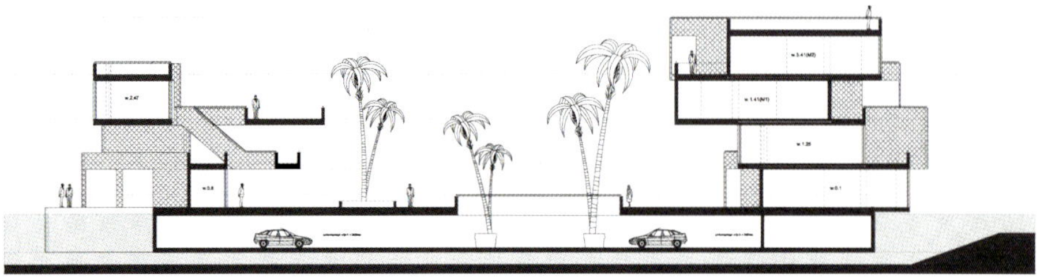

Section Plan 1　剖面图 1

FEATURE 特点分析

Structure

You will not feel any movement. This is because of the dimensions. The building is made of lightweight construction but special care has been given to the lowest possible noise level. The same standards as used for normal land base dwelling.

结构

由于容积的关系，在这里，你将不会感觉到任何的震动。它是轻量化建筑，但却特别注意最大限度地杜绝了噪声，达到了普通陆基建筑的标准。

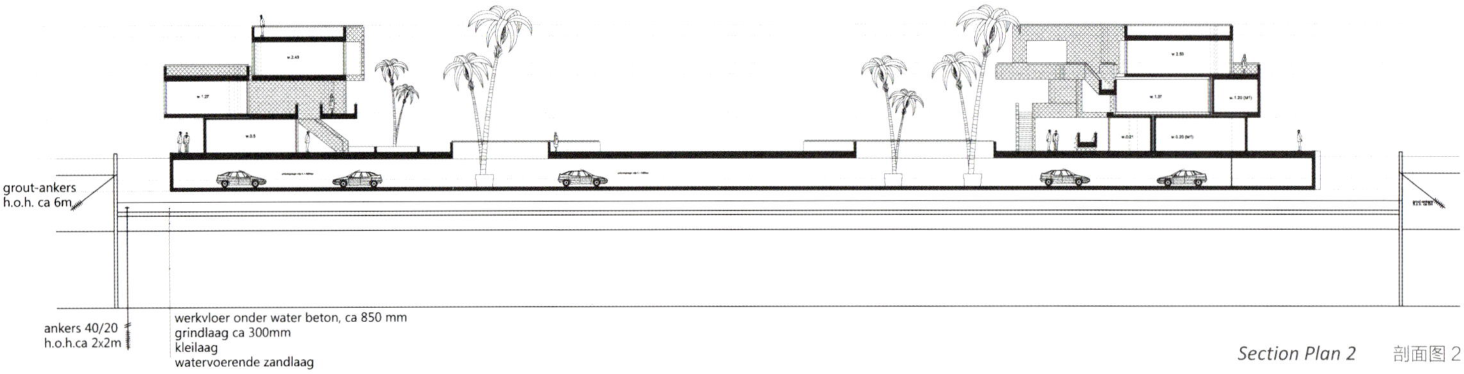

Section Plan 2　剖面图 2

Elevation Rendering Plan　立面效果图

图书在版编目(CIP)数据

突破风格与复制——小高层、多层 / 杨锋　主编.
—南京：江苏人民出版社，2012.3
ISBN 978-7-214-07871-1
Ⅰ. ①突… Ⅱ. ①杨… Ⅲ. ①住宅—建筑设计—世界
Ⅳ. ①TU241

中国版本图书馆CIP数据核字(2011)第280535号

突破风格与复制——小高层、多层　　杨锋　主编

责任编辑　刘　焱　潘绮虹
特约编辑　黄夏炎　付桑林　罗小敏
装帧设计　喻小燕　邵兵兵
策划设计　香港建筑科学出版社
责任监印　彭李君
出　　版　江苏人民出版社(南京湖南路1号A楼　邮编：210009)
发　　行　天津凤凰空间文化传媒有限公司
销售电话　022-87893668
网　　址　http://www.ifengspace.cn
集团地址　凤凰出版传媒集团(南京湖南路1号A楼　邮编：210009)
经　　销　全国新华书店
印　　刷　利丰雅高印刷(深圳)有限公司
开　　本　1020毫米×1440毫米　1/16
印　　张　14.25
字　　数　114千字
版　　次　2012年3月第1版
印　　次　2012年3月第1次印刷
书　　号　ISBN 978-7-214-07871-1
定　　价　208.00元(USD36.00)
(本书若有印装质量问题，请向发行公司调换)